AGRICULTURAL RESOURCES AND ENVIRONMENTAL INDICATORS

AGRICULTURAL RESOURCES AND ENVIRONMENTAL INDICATORS

KEITH WIEBE
AND
NOEL GOLLEHON

Nova Science Publishers, Inc.
New York

For permission to use material from this book please contact us:
Telephone 631-231-7269; Fax 631-231-8175
Web Site: http://www.novapublishers.com

LIBRARY OF CONGRESS CATALOGING-IN-PUBLICATION DATA

Agricultural resources and environmental indicators / Keith Wiebe and Noel Gollehon (editors). -- 2006 ed.
p. cm.
Includes index.
ISBN 13: 978-1-60021-467-7
ISBN 10: 1-60021-467-3
1. Agricultural resources--Economic aspects--United States. 2. Agricultural resources--Environmental aspects--United States. 3. Agriculture--Economic aspects--United States. 4. Agriculture--Environmental aspects--United States. 5. Farms--Valuation--United States. I. Wiebe, Keith Daniel, 1962- II. Gollehon, Noel R. (Noel Ray), 1954-
HD1761.A6223 2006
338.10973--dc22 2006031474

Published by Nova Science Publishers, Inc. ✦ New York

CONTENTS

SUMMARY

Agricultural Resources and Environmental Indicators, 2006 describes trends in resources used in and affected by agricultural production (including natural, produced, and management resources), as well as the economic conditions and policies that influence agricultural resource use and its environmental impacts. Each chapter provides a concise overview of a specific topic with links to sources of additional information.

Keywords: ERS, AREI, agricultural economics, natural resources, land, land use, land values, land ownership, water use, irrigation, water quality, genetic resources, biotechnology, agricultural research, agricultural productivity, global resources, soil conservation, soil erosion, pest management, nutrient management, animal agriculture, organic agriculture, conventional agriculture, conservation policy, land retirement, working lands, wetlands, farmland protection, environmental quality, farm structure, farm ownership, farm management decisions, farm business

ACKNOWLEDGMENTS

Publication of AREI 2006 involved the expert advice and contributions of many people in addition to the authors. We appreciate the guidance and support provided at various stages by Kitty Smith, Mary Bohman, Ralph Heimlich, Robbin Shoemaker, and Joe Cooper. For expert assistance with production, we thank Tom McDonald for overall coordination; Dale Simms for editing; Letricia Womack, Linda Beeler, Pat Vines, and Melody Mathis-Pace for preparation of the html files; Wynnice Napper for preparation of the pdf files; David Nulph for preparing the maps; Susan DeGeorge for assistance with design; and Kathy Kassel, Steve Peterson, and Matt Poyner for assistance with web layout and navigation. For their review comments on all or parts of earlier drafts, we thank Marcel Aillery (ERS), Alex Barbarika (FSA), Charles Barnard (ERS), June Blalock (ARS), John Buckhouse (CSREES), Evert Byington (ARS), Margriet Caswell(ERS), Randy Christopherson (Bureau of Reclamation), Roger Claassen (ERS), Stan Daberkow (ERS), Leslie Deavers (NRCS), Carolyn Dimitri (ERS), Robert Dismukes (ERS), Diane Eckles (NRCS), Ken Erickson (ERS), Mitch Flanagan (NRCS), Jorge Fernandez-Cornejo (ERS), Noel Gollehon (ERS), Ron Green (ARS), LeRoy Hansen (ERS), David Harrington (ERS), John Hassell (Conservation Technology Information Center), Ralph Heimlich (Agricultural Conservation Economics), Paul Heisey (ERS), Mike Hubbs (NRCS), Susan Hutson (U.S. Geological Survey), John Jinkins (FSA), Nigel Key (ERS), C.S. Kim (ERS), John King (ERS), Doug Kleweno (NASS), Doug Lawrence (NRCS), Ruben Lubowski (ERS), Ken Mathews (ERS), William McBride (ERS), Ken Nelson (ERS), Roberta Parry (U.S. Environmental Protection Agency), James Payne (ERS), William Quinby (ERS), Ed Rall (FSA), Marc Ribaudo (ERS), Carmen Sandretto (ERS), Glenn Schaible (ERS), David Schimmelpfennig (ERS), Allan Stoner (ARS), Patrick Sullivan (ERS), Dennis Thompson (NRCS), Marlow Vesterby (ERS), Floyd Wood (NRCS), and Jet Yee (ERS). (Acronyms represent agencies within USDA.)

LIST OF CONTRIBUTING AUTHORS

Marcel Aillery
Eldon Ball
Charles Barnard
Shawn Bucholtz
Margriet Caswell
Roger Claassen
Stan Daberkow
Kelly Day-Rubenstein
Robert Dubman
Jorge Fernandez-Cornejo
Linda Foreman
Dwight Gadsby
Noel Gollehon
Catherine Greene
LeRoy Hansen
Paul Heisey
Daniel Heller-stein
Robert Hoppe
Wen Huang
Rob Johansson
Jim Johnson
C.S. Kim
Janet Livezey
Michael Livingston
Ruben Lubowski
William McBride
Mitchell Morehart
Cynthia Nickerson
Craig Osteen
James Payne
William Quinby
Marc Ribaudo
Carmen Sandretto

Glenn Schaible
Marlow Vesterby
Keith Wiebe

PREFACE

WHAT IS THE ISSUE?

Agricultural production both depends on and influences a wide range of natural and other resources. These include land, water and genetic material as well as knowledge, production technologies and management skills. Concise and accurate information on these resources can help public and private decisionmakers better understand the complex interactions between public policies, economic conditions, farming practices, conservation and the environment.

WHAT DID WE DO?

Agricultural Resources and Environmental Indicators, 2005 describes patterns and trends in land, water, biological resources, management skills and commercial input use; reports on the condition of natural and other resources used in the agricultural sector; and describes public policies and programs as well as economic factors that affect resource use, conservation and environmental quality in agriculture. Each chapter synthesizes, updates, and provides links to more detailed information available in ERS reports, data products and briefing rooms on the ERS website. Three previous editions of *AREI* (1994, 1997 and 2003) are also available on the ERS website.

WHAT DID WE FIND?

Agricultural resource use depends on the decisions made by the operators of the nation's 2.1 million farms, which are shaped in turn by market conditions, public policies, and the specific characteristics of individual farms and households. When making these decisions, farm operators have clear incentives to consider the impacts on their own well-being and that of their households, but weaker incentives to consider impacts that occur farther away. This raises ongoing challenges in managing the nation's agricultural resources and motivates ongoing efforts to balance public and private goals. Among our findings:

- Land continues to shift between agriculture and other uses. Cropland has declined but losses do not threaten the nation's capacity to produce food and fiber.

- Competition for water is increasing, but potential remains to increase agricultural water conservation through improved irrigation technology and management.
- Increasing concentration in animal production can have adverse impacts on air and water quality. A variety of voluntary and regulatory measures have been introduced at Federal, State and local levels to mitigate these impacts.
- Public and private agricultural research and development (including advances in biotechnology) have helped drive rapid growth in agricultural productivity, but public R&D investment and productivity growth have slowed in recent years.
- Most farms are operated by a single operator or an operator and spouse, but most production comes from farms with larger and more complex management teams.
- Soil erosion declined by more than a billion tons per year between 1982 and 1997, a quarter of which can be attributed to conservation compliance requirements.
- Use of commercial fertilizers and pesticides has been steady or declining in recent years, due to improvements in technology and other factors.
- Certified organic farmland more than doubled between 1992 and 2003, and USDA national standards for organic production and processing came into effect in 2002.
- The Farm Security and Rural Investment Act of 2002 sharply increased funding for conservation programs. Land retirement remains a key strategy, but much of the increase focused on programs for working cropland and grazing land.
- Improved information on natural, produced, and management resources used in agriculture can help public and private decisionmakers better understand the complex interactions between public policies, economic conditions, farming practices, conservation and the environment.

PART 1: LAND AND FARM RESOURCES

In: Agricultural Resources and Environmental Indicators
Editor: Keith Wiebe and Noel Gollehon, pp. 3-10
ISBN 978-1-60021-467-7

Chapter 1

LAND USE

Ruben Lubowski, Marlow Vesterby and Shawn Bucholtz

The three major uses of land in the 48 contiguous States are grassland pasture and range, forest-use land, and cropland, in that order. Total cropland (used for crops, used for pasture, and idled) declined 6 percent over 1969-2002. Farm policy changes have reduced the acreage idled under Federal programs since 1996.

INTRODUCTION

Land-use changes can affect the environment and the sustainability of production. Because impacts on the environment—including erosion, water quality, and wildlife habitat—are typically not reflected in private profit calculations, land-use choices that are optimal for an individual may not be optimal for society. This difference suggests the possibility of public policies that more closely align land-use decisions with social objectives.

The allocation of a fixed land base among competing uses is determined by the relative returns to the different uses, which vary according to land quality and location. A landowner seeking to maximize profits will allocate a land parcel to the use that yields the highest expected economic return, after the costs of conversion. As relative returns change along with market conditions, technological advancements, or government policies, land-use patterns tend to adjust accordingly (see the "Land Use, Value and Management" Briefing Room on the ERS website).

Land-use change is dynamic. With the exception of urban land, changes occur to and from major land uses. For example, 44 million acres left the cropland and pasture category from 1992 to 1997 while 21 million acres shifted into the category, resulting in a net loss of 23 million acres (USDA/NRCS, 2000).

MAJOR LAND USES IN THE UNITED STATES

Major Land Uses is a land-use inventory conducted periodically by ERS. This series contains acreage estimates of major uses by region and State, coinciding with each census of agriculture from 1945 through 2002. (See the glossary for detailed definitions of the major land uses.)

Because Alaska and Hawaii have very little crop area, we focus on the contiguous 48 States. The total land area of the 48 contiguous States is approximately 1.9 billion acres, with an additional 365 million acres in Alaska and a little over 4 million acres in Hawaii (Table 1).

Table 1. Major uses of land, United States, 2002[1]

Land use	48 States	United States	48 States	United States
	Million acres		Percent of total	
Cropland[2]	441	442	23.3	19.5
Grassland pasture and range	584	587	30.8	25.9
Forest-use land[3]	559	651	29.5	28.8
Special uses[4]	153	297	8.1	13.1
Urban	59	60	3.1	2.6
Miscellaneous other land	97	228	5.1	10.1
Total land area[5]	1,894	2,264	100.0	100.0

[1] See the "Major Land Uses" data product on the ERS website for estimates of major uses by region and State, coinciding with each census of agriculture, from 1945 through 2002.

[2] All land in the crop rotation (used for crops, used for pasture, idle cropland). Includes about 34 million acres idled under the Conservation Reserve Program.

[3] Total forest land as classified by the U.S. Forest Service minus an estimated 98 million acres of forested land used for parks, wildlife areas, and other special uses.

[4] Rural transportation areas, land used primarily for recreation and wildlife purposes, various public installations and facilities, farmsteads, and farm roads/lanes. Excludes urban land in contrast to Major Land Uses, Aggregate Data.

[5] Distributions by major use may not add to totals due to rounding.

Sources: USDA/ERS based primarily on reports and records of the Census Bureau and Federal, State, and local land management and conservation agencies. See the Major Land Report (Lubowski et al., 2006) for information about the 2002 land-use estimates.

Grassland pasture and range, the largest use of land, accounted for 584 million acres (31 percent) of the 48 States in 2002 (Table 1). This compares with 636 million acres in the mid-1960s. Due to improvements in the forage quality and productivity of grazing lands, less pasture and range is needed to sustain grazing herds. The inventory of domestic animals, particularly sheep, has also been declining in recent years, further reducing pasture/range demand (USDA/NASS, 2004).

Forest-use land, the second largest major use, declined from about 32 percent of total land in 1945 to about 30 percent in 2002. A broader category, all land with forest cover, comprised 33 percent of the land base in 2002 (Smith et al., 2004). While forest-use land increased 1 percent between 1997 and 2002, it declined from 612 million acres in 1964 to 559 million acres in 2002. Much forest-covered land is in "special uses" (parks, wilderness areas, and wildlife areas) that prohibit forestry uses such as timber production. Forested land in these special uses increased from 23 million acres in 1945 to about 98 million acres in 2002.

Cropland comprises the third largest use of land, covering 23 percent of the contiguous States in 2002 (Table 1). Since 1945, cropland ranged from a high of 478 million acres in 1949 to a low of 441 million acres in 2002. Total cropland has trended downward since the late 1960s, and decreased by 13 million acres (3 percent) from 1997 to 2002.

The total cropland base includes cropland used for crops, cropland used for pasture, and cropland idled. These components vary more than total cropland. Since 1945, the amount of cropland used for crops has ranged from as much as 383 million acres in 1949 and 1982 to a minimum of 331 million acres in 1987. Total acreage used for crops exhibited two major cycles between 1945 and 1987, with cropland moving from idle to crop use and back again. Cropland used for crops increased from 331 to 349 million acres over 1987-97, and then declined to 340 million acres in 2002, about 5 percent below the average acreage for 1910-97. Since 1945, cropland used for pasture varied from 47 million acres in 1945 to 88 million acres in 1969.

Special uses include rural transportation; rural parks and wildlife; defense and industrial uses; and farmstead, farm roads/lanes, and other on-farm uses. These special uses increased from 85 million acres (4 percent of the land area of the contiguous 48 States) in 1945 to 153 million acres (8 percent) in 2002.

Land in transportation uses (highways and roads, railroads, and airports in rural areas) increased by 4 million acres (17 percent) between 1945 and 1982. Transportation uses declined by about 0.5 million acres from 1982 to 1992 due to the abandonment of railroad facilities and rural roads, and the classification of some transportation uses as urban areas.

Land used for recreation and wildlife areas (Federal and State parks, wilderness areas, and wildlife refuges) expanded 344 percent from 1945 to 2002 (an increase of 78 million acres). The increase came mostly from conversion of Federal lands, previously in forest and grassland pasture and range. Land in defense and industrial uses declined by 10 million acres (40 percent) from 1945 to 2002. Farmsteads, farm roads, and other farm uses declined by 4 million acres (29 percent) between 1945 and 1997. This decline reflects trends toward fewer farms and larger, more consolidated farms, as well as an increasing tendency for farm households to live off the farm.

In response to expanding U.S. population, land in urban uses—including homes, schools, office buildings, shopping sites, and other commercial/industrial uses—increased from 15 million acres in 1945 to 25 million acres in 1960, 47 million acres in 1980, and 59 million acres in 2002. While the U.S. population nearly doubled, the amount of land urbanized quadrupled. However, urban uses still comprise only 3 percent of the total land area of the contiguous States.

Miscellaneous other land uses decreased from 1945 to 1964, and have since trended upward, showing a 54-percent increase from 1964 to 2002 reflecting improved data and reclassification of grazing and forest lands. These uses include marshes and open swamps not included in other major land uses, bare rock areas, deserts, some rural residential areas, and other uses not inventoried. Wetlands are defined by soil and hydrological characteristics and may occur on land in many different uses (see Chapter 6, "Wetland Status and Trends").

REGIONAL CHANGES IN LAND USE

While land in every use occurs in all 10 regions of the contiguous States, some uses are more concentrated in some regions than in others (Figure 1). Regions with the largest cropland acreage are the Northern Plains, Corn Belt, and Southern Plains. Grassland pasture and range is concentrated in the Mountain and Southern Plains regions. Acreage in forest use, special and miscellaneous other uses is highest in the Mountain region.

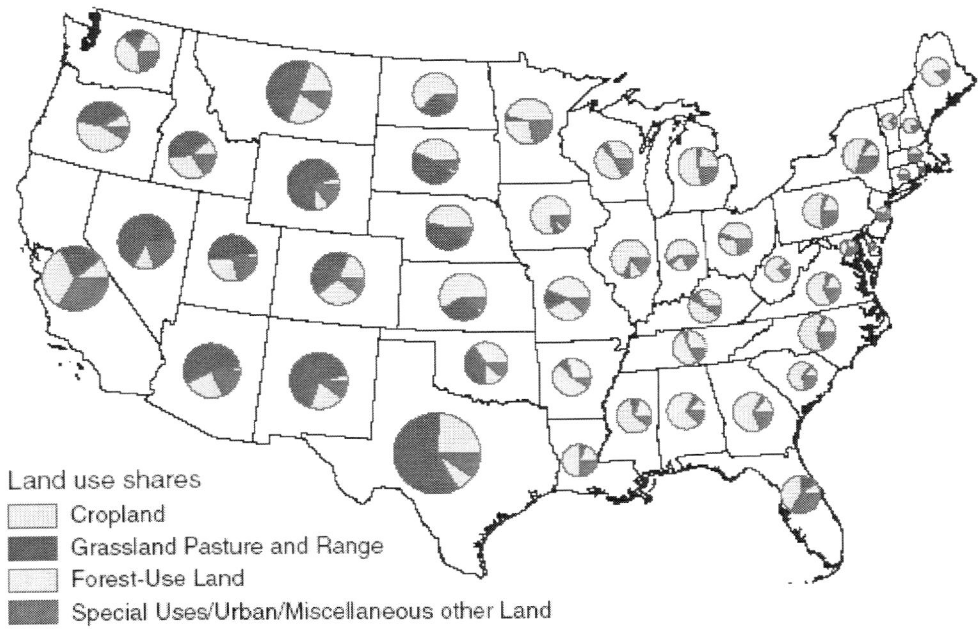

Source: USDA/ERS, Major Land Uses Data Product.

Figure 1. Shares of land in major uses, 48 contiguous States, 2002.

The Northeast, Appalachian, Southeast, Delta States, and Lake States regions lost cropland between 1945 and 2002. The largest increases occurred in the Northern Plains and Mountain regions, with smaller increases in the Corn Belt, Southern Plains, and Pacific regions. Western increases may have resulted in part from federally subsidized irrigation water (see Chapter 4, "Irrigation Resources and Water Costs").

Nine of the 10 regions lost grassland pasture and range between 1945 and 2002. While grassland pasture and range increased 11 million acres (10 percent) in the Southern Plains, the Northeast region lost about 70 percent of its grassland pasture and range, and the Appalachian and Lake States regions lost more than 50 percent. The Northeast and Appalachian regions saw the reforestation of grassland, loss of some grassland to urbanization, and concentration of the dairy industry. Decreases in the Corn Belt, Northern Plains, and Mountain regions were likely associated with the conversion of some grassland pasture and range to cropland.

CROPLAND USE AND FEDERAL PROGRAMS

While total cropland acreage has varied up and down and generally declined since 1969, greater shifts have occurred between cropland used for crops and cropland idled, mostly because of Federal programs. Cropland used for pasture has exhibited less variation in acreage than cropland idled.

Most cropland used for crops is harvested, but typically 2-3 percent experiences crop failure and 5-10 percent is cultivated summer fallow (Table 2). In 2002, farmers harvested one or more crops on an estimated 307 million acres of cropland, down 4 percent from 1997. About 8 million acres of the total harvested were double-cropped. When double-cropped land is counted twice, total acres harvested rise to 315 million acres.

Table 2. Major uses of cropland, 48 contiguous States, 1992-2004

Cropland	1992	1997	2002	2004[1]
	Million acres			
Cropland used for crops[2]	337	349	340	336
Cropland harvested[3]	305	321	307	312
Crop failure	8	7	17	9
Cultivated summer fallow	24	21	16	15
Cropland idled by all Federal programs[2]	55	33	34	35
Annual programs	19	0	0	0
Conservation Reserve Program	35	33	34	35
Total, specified uses[2,4]	392	382	374	372

[1] Preliminary, subject to revision.
[2] Breakdown may not add to totals due to rounding.
[3] A double-cropped acre is counted as 1 acre.
[4] Does not include cropland pasture or idle land not in Federal programs that is normally included in the total cropland base.

Sources: USDA/ERS, based on USDA/ERS, 1997a; USDA/NASS, 1998, 1999a, 2000a; 2004a; 2005; and unpublished data from USDA/FSA and USDA/NASS.

Cropland used for crops was at a record high of 387 million acres in 1949, when no acres were idled by Federal programs. In 1972, cropland used for crops was near a record low of 334 million acres as Federal programs idled 61 million acres. Cropland used for crops climbed to 387 million acres in 1981, when Federal programs idled no cropland, and dropped to 333 million in 1983, when Federal program set-asides reached a historic peak of 78 million acres under the Payment-in-Kind (PIK) program. The Federal Agricultural Improvement Act (FAIR) of 1996 eliminated all Federal acreage reduction programs other than the Conservation Reserve Program (CRP) (see Chapter 22, "Land Retirement Programs"). Between 1983 and 2002, cropland used for crops increased overall, while total acreage idled by Federal programs decreased to 34 million acres. From 1997 to 2002, cropland used for crops declined from 349 to 340 million acres, while acreage idled under CRP increased by about 1 million acres. Between 2002 and 2004, acreage in CRP increased by an additional 1 million acres, while cropland used for crops declined by about 4 million acres (Table 2).

The 14-million-acre drop in harvested cropland between 1997 and 2002 was coincident with a decrease in cultivated summer fallow and an increase in failed acres due to widespread drought.

Crop failure occurred on 17 million acres, over 5 percent of the acreage planted, in 2002. This failed acreage was the largest since 1956. The use of summer fallow has been decreasing since the late 1960s, and stood at 15 million acres in 2002, down from 42 million acres in 1969.

Four crops—corn for grain, soybeans, wheat, and hay—accounted for 80 percent of all crop acres harvested in 2002 (Figure 2). The additional 17 "principal" crops accounted for another 15 percent of harvested area. Vegetables, fruits, nuts, melons, and all other crops accounted for 4.5 percent of crop area harvested in 2002.

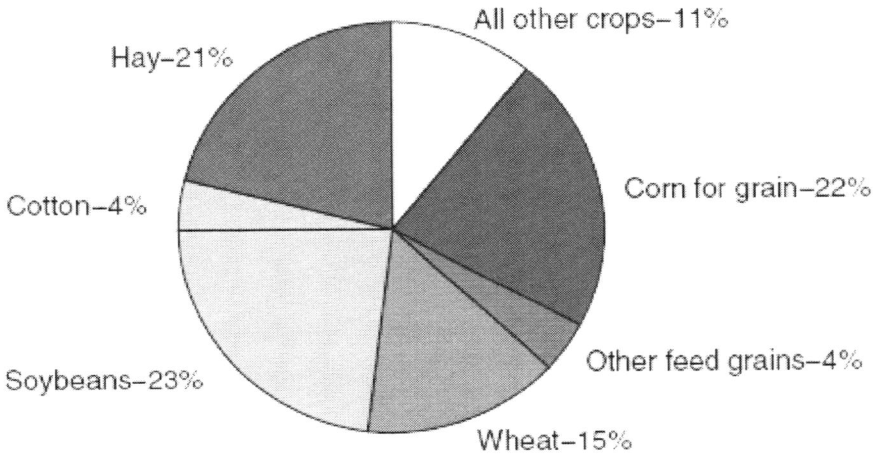

Source: USDA/ERS, based on USDA/NASS, 1999a, 1999b, 1999c, 1998, 2004.

Figure 2. Harvested crops in the 48 contiguous States, 2002.

URBANIZATION OF AGRICULTURAL AND OTHER RURAL LAND

Cropland conversion to urban uses is largely irreversible, so it is important to know the rate of conversion and how much of the loss is replaced from other land uses (see briefing room "Land Use, Value and Management" on the ERS website). Excessive loss of cropland to urban uses could lessen the production of food and fiber and the supply of rural amenities, such as open space, watershed protection, and rural lifestyles. A variety of Federal, State, local, and private programs address such concerns (see Chapter 26, "Farmland Protection Programs").

Although urban land constituted less than 3 percent of the U.S. land area in 2000, 79 percent of the population lived there (Table 3). Even large percentage increases in urban area would amount to small decreases in rural area since it is so vast. The rate of expansion (by decade) of urban area has declined from 39 percent during the 1950s, to about 36 percent during the 1960s and the 1970s, and to 18 percent in the 1980s. According to the Census

Bureau, urban area was 59 million acres in 2000, just 7 percent above the previous estimate for 1990 (DOC/BOC, 2002). However, the Census Bureau adopted a new definition of urban area for the 2000 Census, improving the precision of urban area measurement but making it more difficult to compare urban area before and after this year. If urban area for 1990 is recalculated using the 2000 definition, the Census Bureau's estimate falls to 51 million acres, implying a 15-percent increase in urban area from 1990 to 2000.

The National Resources Inventory (NRI), conducted by USDA's Natural Resources Conservation Service (NRCS) in cooperation with Iowa State University, is an alternative source for estimates of urban and rural areas. The NRI uses a consistent definition for built-up areas, though it differs from the definition used by the Census Bureau. According to the NRI, "developed land," which includes large and small urban and built-up areas as well as rural transportation land, totaled 107 million acres in 2002 in the contiguous United States. The NRI indicates that developed land increased by 14 million acres (19 percent) over 1982-92 and 21 million acres (24 percent) over 1992-2002 (USDA/NRCS, 2004).

Table 3. U.S. population and urban area, 1950-2000

Year	U.S. population Total	Urban	Portion urban	Urban area[1]	Urban area increase[2]
	Millions		Percent	Million acres	Percent
1950	151	97	64	18	
1960	179	125	70	25	39
1970	203	150	74	35	36
1980	227	167	74	47	37
1990	249	187	75	56	18
2000[3]	281	222	79	59	1

[1] Data differ from table 1.1.1 due to different data sources and time periods.
[2] Percent increase over urban area 10 years earlier.
[3] The 2000 urban area estimates are not directly comparable to estimates in prior years due a change in the definition of urban areas in the 2000 Census of Population and Housing. The relatively small change in urban area between 1990 and 2000 should be viewed as a result of this definitional change, rather than as a reflection of a slowing rate of urbanization.
Sources: USDA/ERS, based on DOC/BOC, 2002, 1999; and Frey, 1983.

Land converted to urban uses comes from several different major land uses. The NRI indicates that 20 percent of new developed land came from crop-land between 1997 and 2002. Prime cropland—land that has the best combination of physical/chemical characteristics for agricultural production—is converted to developed uses at about the same rate (5 percent per year) as nonprime cropland (USDA/NRCS, 2003). About 21 percent of rural nonFederal land is prime and 26 percent of crop land converted to urban uses over 1997-2001 was prime.

Rural land, defined as all land that is not urban, contains rural residential land, consisting of houses and associated lots. Non-farm rural residential area was estimated to be about 94 million acres in 2002, up from 56 million acres in 1980. The average rate of increase in rural residential land was 1.7 million acres per year from 1980 to 2002. Combining both rural and urban residential land, the total increase in residential area was about 2 million acres per year during this period.

REFERENCES

[1] Lubowski, Ruben N., Marlow Vesterby, Shawn Bucholtz, Alba Baez, and Michael J. Roberts (2006). *Major Uses of Land in the United States, 2002.* EIB-14. U.S. Dept. Agr., Econ Res. Serv., May.

[2] Smith, Brad W., Patrick D. Miles, John S. Vissage, and Scott A. Pugh (2004). *Forest Resources of the United States, 2002: A Technical Document Supporting the USDA Forest Service 2005 Update of the RPA Assessment.* General Technical Report NC-241. Forest Service, North Central Research Station, U.S. Department of Agriculture, St. Paul, MN.

[3] U.S. Department of Agriculture, National Agricultural Statistics Service (NASS) (2004). *2002 Census of Agriculture.* Vol. 1: Part 51, Chapter 2, AC-02-A-51, United States Summary and State Data. June.

[4] U.S. Department of Agriculture, Natural Resources Conservation Service (NRCS), and Iowa State University, Center for Survey Statistics and Methodology (2003). *National Resources Inventory: 2001 Annual NRI.* July.

[5] U.S. Department of Agriculture, Natural Resources Conservation Service (NRCS), and Iowa State University, Center for Survey Statistics and Methodology (2000). *Summary Report, 1997 National Resources Inventory (revised December 2000)*, including associated database files. Dec.

[6] U.S. Department of Commerce, Bureau of the Census (BOC) (2002). *2000 Census of Population and Housing, Summary Population and Housing Characteristics, United States.* PHC-1-1. Nov.

In: Agricultural Resources and Environmental Indicators
Editor: Keith Wiebe and Noel Gollehon, pp. 11-17
ISBN 978-1-60021-467-7

Chapter 2

FARM REAL ESTATE VALUES

Charles Barnard

Farm real estate values and cash rents are important indicators of the financial condition of the farm sector. Real estate comprises a substantial share of the asset portfolio of farm households. Farm real estate values are influenced by net returns from agricultural production, capital investment in farm structures, interest rates, government commodity programs, property taxes, and non-farm demands for farmland. Values have been steadily rising since 1987, but the inflation-adjusted (real) value of U.S. farm real estate is still below its 1982 peak. Cash rents have also been increasing in recent years.

Farmland values rose rapidly during the 1970s and early 1980s, followed by a sharp decline during 1982-87, then a slow upward trend beginning in 1987 (Figure 3). Since 1987, average farmland values in the Nation have increased 127 percent, from $599 per acre to $1,360 in January 2004. In real or inflation-adjusted terms (GDP deflator), however, this amounts to a 53-percent gain. It was not until January 1, 1995, that the average nominal value per acre surpassed the record high of $823 set in 1982. But the January 2004 average is still 8 percent below the 1982 average on a real (or inflation-adjusted) basis.

The 7.1-percent nominal increase in the national average value of agricultural real estate during 2004 marked the 17th consecutive increase since 1987. Over the previous 4 years, in particular, farm real estate values had increased substantially in all U.S. regions (Table 4). Most notable is a 42-percent increase in the Lake States, versus 25 percent for all regions combined.

The increases were widespread, with most States exhibiting increases for farm real estate, cropland, and pasture. As of January 2004, several Northeast States continued to record the highest average per-acre values for farm real estate, with Connecticut and Rhode Island exceeding $10,000 per acre (Figure 4). These values reflect continued pressure from nonagricultural sources for conversion to residential or other urban-related uses. The high values in States such as California, Florida, Ohio, Illinois, Indiana, and North Carolina are the consequence of urban pressures, the production of high-value crops, or high soil fertility. The low real estate values for many States in the Northern Plains, Southern Plains, and Mountain regions can be attributed to large amounts of arid rangeland and less productive cropland. New Mexico, Wyoming, and Montana recorded the lowest average values per acre.

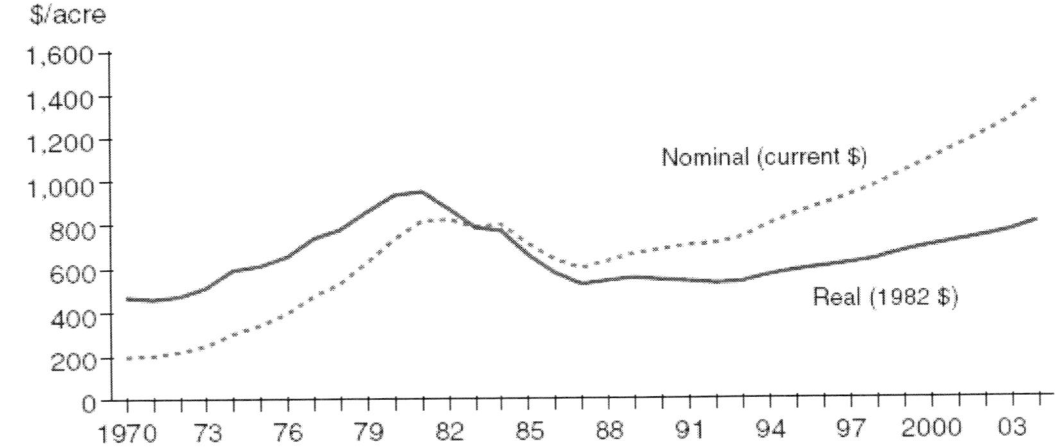

Source: USDA, Economic Research Service.

Figure 3. Average real and nominal values of U.S. farm real estate.

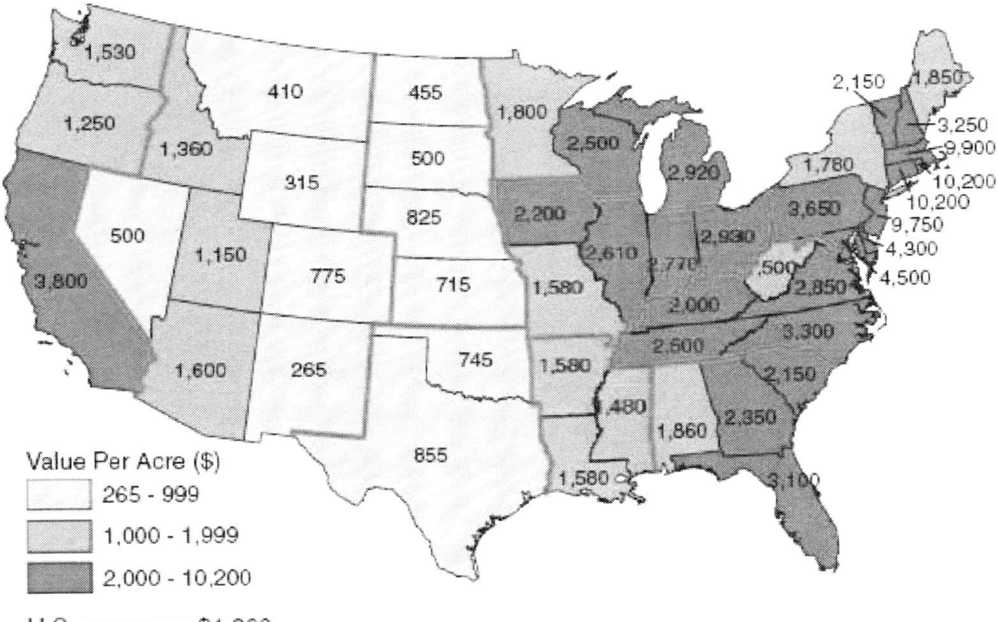

Source: USDA, NASS, Sp Sy 3 (04), August 2004.

Figure 4. Farm real estate value per acre, January 1, 2004.

Table 4. Farm real estate values, by farm production region, January 1 for selected years

Region	1982	1987	2000	2001	2002	2003	2004
				$/acre			
Northeast	1,367	1,491	2,660	2,830	3,000	3,200	3,400
Lake States	1,234	707	1,560	1,700	1,870	2,010	2,220
Corn Belt	1,642	900	1,890	1,950	2,030	2,130	2,300
Northern Plains	547	331	535	556	576	594	632
Appalachian	1,083	1,004	1,990	2,120	2,250	2,370	2,500
Southeast	1,095	1,055	1,920	2,030	2,140	2,270	2,420
Delta States	1,135	757	1,270	1,330	1,390	1,490	1,550
Southern Plains	576	532	672	715	755	788	832
Mountain	325	257	448	471	500	523	550
Pacific	1,346	1,084	2,000	2,120	2,240	2,350	2,480
48 States	823	599	1,090	1,150	1,210	1,270	1,360

Source: National Agricultural Statistics Service, USDA.

Cash Rents

Nearly a third of U.S. farmland is operated under some form of lease, according to the 2002 Census of Agriculture. The most common form of lease, the cash rental agreement, is a fixed payment negotiated before planting. Share rental agreements, by contrast, vary with the amount of product harvested. Under cash rental arrangements, the tenant bears all of the production and market-price risk; share rental arrangements divide production and market risks between tenant and landlord.

Cash rents are generally considered a short-term indicator of the return to a landowner's investment. To tenants, though, cash rents are a major production expense and, like farm real estate values, have been increasing for a number of years (Figure 5).

Because rents reflect the income-earning capacity of the land, they vary widely across the country. Cropland rents tend to be highest in areas where higher-value crops are grown. The highest average cash rents in 2004 were reported for irrigated land in California, at $300 per acre (Figure 6). California produces large quantities of high-value specialty crops, vegetables, fruits, and nuts. Cropland most suitable for corn and soybean production, principally in the Midwest, also commands high rents. The highest rents for nonirrigated cropland in 2004 were reported as $126 per acre in both Illinois and Iowa (Figure 6).

During 2004, average cash rents for pasture varied from $37 per acre in Wisconsin to $1.70 per acre in New Mexico. States in the Appalachian, Delta, Southern Plains, and Pacific regions uniformly recorded increases from 2003.

Grazing Fees

Grazing fees for use of pasture or rangeland are also a form of cash rent, except that payment is based on "grazing units" rather than tracts of land (acres). A grazing unit is defined on an animal-unit-month (AUM) basis, which is one cow or cow-calf pair, or seven sheep/goats, feeding for 1 month (NASS, 2005). Grazing fees on public lands administered by the

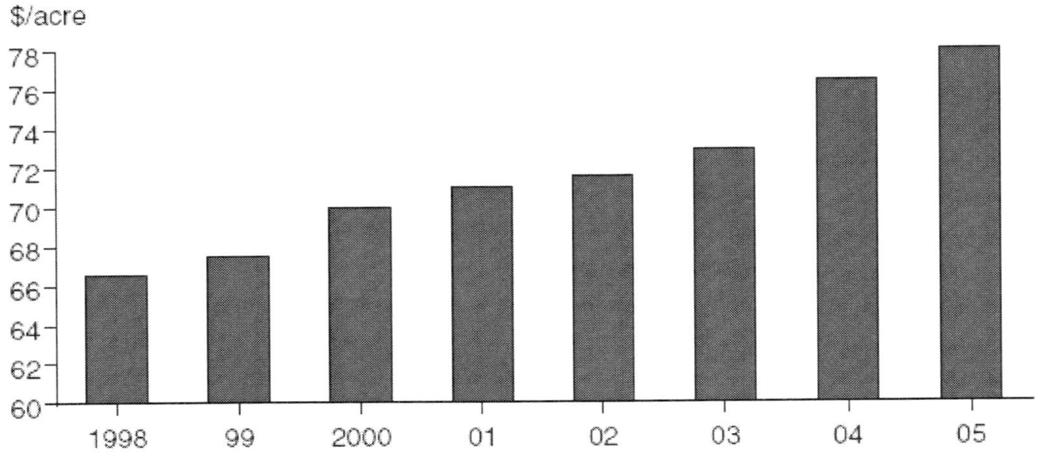

Source: USDA, NASS, August 2005.

Figure 5. U.S. average cropland rent, nominal dollars per acre, 1998-2005.

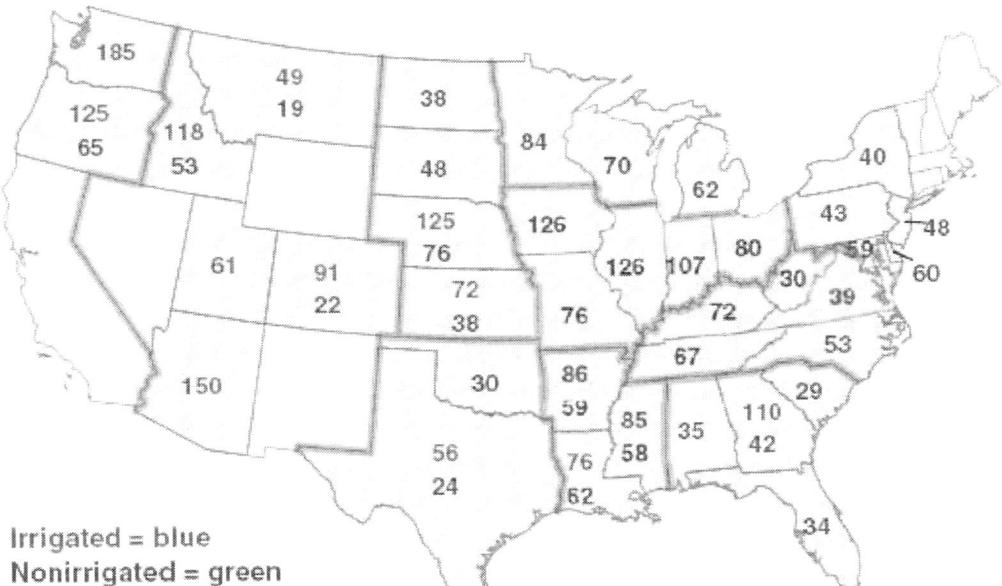

Source: USDA, NASS, August 2004.

Figure 6. Average per-acre cash rent for irrigated and non-irrigated cropland, 2004, selected States.

Bureau of Land Management (BLM) of the U.S. Department of the Interior, and the Forest Service (FS) of the U.S. Department of Agriculture (USDA) are set by law. These fees vary annually according to a legislated formula, which links the fees to changes in the cost of production. As a result of the formula, 2005 grazing fees on public land were set at $1.79 per

AUM. That marks the second consecutive year in which grazing fees were set above the statutory minimum $1.35 per AUM.

Grazing rates on privately owned nonirrigated land in 16 Western States averaged $14.30 per AUM in 2004. Rates ranged from $7 in Arizona and Oklahoma to $23 per AUM in Nebraska. Private grazing rates have trended upward since the early 1990s.

Factors Affecting Farm Real Estate Values

Traditionally, farmland value was based on its agricultural productivity. Particularly in the more rural areas of the Nation, where farmers still account for most farmland purchases, net returns to agricultural uses are the principal determinant of farmland value. Interest rates, capital investment in farm structures, and many other factors also influence productivity and thus the agricultural value of farm real estate. But today, many factors unrelated to productivity—including urban influence, government program payments, and rural amenities—contribute to the value of land in rural areas. In fact, these factors may be more important than productivity. High levels of direct government payments, which have occurred particularly since 1999, may have influenced farmland values in some regions.

Urban Influence

Farmland near cities has seen its value inflated by demand for conversion to non-farm uses. As the U.S. population continues to grow and disperse, even primarily rural States such as Iowa are experiencing urban-related influences on farmland values. Commuters, who can now travel farther or even telecommute, are often willing to pay more than agricultural value in order to live in primarily rural areas. Other families develop hobby farms, second homes, or recreational structures in rural areas. In Iowa, for instance, there are now more non-farmers living in rural areas than there are farmers (see Chapter 1, "Land Use"). Other nonagricultural factors that may contribute value include the potential to concurrently use farmland for fee-based hunting, fee-based recreation, or wildlife viewing.

Non-farm influences on agricultural real estate values have gained increased attention as interest in farmland preservation, suburban "sprawl," and habitat conservation has grown. Recent research indicates that non-farm influence accounts for 25 percent of the market value of U.S. farmland (Barnard, 2000). An ERS report recently addressed issues surrounding development of new houses, roads, and commercial buildings at the fringe of existing urban areas. This "sprawl" into the countryside can intersperse sometimes incompatible urban-related development with existing agriculture. Metropolitan Statistical Areas (MSAs) contain 20 percent of U.S. land area and 80 percent of U.S. population (Bureau of the Census, 2000). The area also contained more than a third of all U.S. farms in 2003 and produced about a third of agricultural production value.

Direct Federal Payments

An array of government policies influence the income derived from farmland, and hence its value. Federal commodity and conservation programs are the most obvious. But also important are farm credit programs, State and local zoning regulations, habitat and species protection laws, infrastructure development (such as roads and dams), environmental regulations, and even property and income tax policy.

Previous research has shown that capitalization of expected payments increases cropland values (Barnard et al., 2001). Also, the degree to which direct Federal payments are capitalized into cropland values depends upon the issuing program (Goodwin and Mishra, 2003). If direct payments are capitalized into cropland values, as many theorize and some research has demonstrated, then a reduction in payments could signal a decline in crop-land values and a loss of wealth for landowners. Further, ERS estimates that the degree to which direct payments (even from the same program) are capitalized into cropland values varies widely, with capitalization greatest in the Northern Plains. So from a policy perspective, the effect of program changes on cropland values would vary depending on the dominant program crop in a region.

Other Market-Related Factors

Interest rates, particularly inflation-adjusted ones, are especially important determinants of U.S. farmland values. As proxies for the discount rate, interest rates determine the current value of expected future earnings from land: for a given pattern of future earnings, higher (lower) interest rates imply lower (higher) land values. During much of the mid- to late 1970s, real interest rates were actually negative, providing a strong incentive to borrow money. Some of the borrowed money was used to purchase rapidly appreciating farmland. Conversely, real interest rates jumped from 1981 to 1985 when nominal interest rates increased rapidly just as expectations of future inflation were diminishing. The resulting increase in the real interest rate of mortgages has been cited as a cause of the slide in farmland values in the early and mid-1980s.

Inflation, lending policies of farm credit agencies and banks, and speculation also affect farmland values. And of course farmland values vary by site-specific characteristics like access to major highways, proximity to commodity and input markets, aesthetic appeal, and homesite potential.

Nonmarket Public Goods Provided by Farmland

Farmland also provides nonmonetary benefits. Until recently, these "rural amenity" benefits were supplied in such abundance that they were rarely acknowledged. But as the Nation becomes more urbanized, with the concomitant loss of farms and interspersion of urban-related activities, the decrease in those amenities has become a source of concern. The nonmonetary benefits potentially reduced or eliminated by loss of farmland and open space include recreation opportunities, aesthetic enjoyment from viewing landscapes and wildlife, environmental quality, and nostalgia related to the historic and cultural significance of rural life. It is these "rural amenity" benefits that many farmland preservation programs seek to protect. A more extended discussion is available in Chapter 5, "Farmland Protection Programs", an ERS report on Rural Amenities (McGranahan, 1999), and current ERS activities examining farmland.

REFERENCES

[1] American Farmland Trust (1997). Saving American Farmland: What Works. Northampton, MA.

[2] Barnard, C.H., R. Nehring, J.T. Ryan, and R. Collender (2001). "Higher Cropland Value from Farm Program Payments: Who Gains?" Agricultural Outlook, U.S. Dept. of Agr., Econ. Res. Serv., AGO-286, Nov.

[3] Barnard, Charles H. (2000). "Urbanization Affects a Large Share of Farmland," Rural Conditions and Trends, U.S. Dept. of Agr., Econ. Res. Serv., Vol. 10, No. 2.

[4] Federal Reserve Bank of Kansas City (2000). Survey of Agricultural Credit Conditions. Dec. 31.

[5] Goodwin, Barry, and Ashok Mishra (2005). "Another Look at Decoupling: Do Direct Payments Affect Production Decisions." 2005 AAEA annual meetings, Providence, RI.

[6] Heimlich, Ralph E., and William D. Anderson (2001). Development at the Urban Fringe and Beyond: Impacts on Agriculture and Rural Lands. U.S. Dept. of Agr., Econ. Res. Serv., AER-803, June.

[7] McGranahan, David A. (1999). Natural Amenities Drive Rural Population Change. U. S. Dept. of Agr., Econ. Res. Serv., AER-781, Sept.

[8] U.S. Bureau of the Census, U.S. Dept. of Commerce (2000). Geographic Areas Reference Manual, Chapter 12.

[9] U.S. Department of Agriculture, National Agricultural Statistics Service (2000). Agricultural Prices: January 2005 and earlier issues.

In: Agricultural Resources and Environmental Indicators ISBN 978-1-60021-467-7
Editor: Keith Wiebe and Noel Gollehon, pp. 19-26

Chapter 3

LAND OWNERSHIP AND FARM STRUCTURE

Robert Hoppe

Small family farms account for most land owned by farms, making them important to conservation. Leased land is a large share of farm operations, and farmers' tenure affects their use of conservation measures, particularly measures with a long payback period. The trend of concentrating livestock on fewer acres than in the past raises environmental concerns.

OWNERSHIP OF U.S. LAND

The land surface of the United States covers 2.3 billion acres. Private owners held 61 percent in 2002, the Federal Government 28 percent, State and local governments 9 percent, and Indian reservations 3 percent (Figure 7). Virtually all cropland is privately owned, as is three-fifths of grassland pasture and range and over half of forestland. Federal, State, and local government holdings consist primarily of forestland, rangeland, and other land. Most land in Federal ownership—largely in the West—is managed by the Department of the Interior (68 percent) and the Department of Agriculture (28 percent) (U.S. GSA, 2005). (For more information, see Chapter 1, "Land Use".)

Farm operators do not own all the land used in agriculture. According to the 1999 Agricultural Economics and Land Ownership Survey (AELOS), farmers held 58 percent of the land in farms in 1999 (USDA, 2001). These landowning farmers also made up 58 percent of the 3.4 million farmland owners.

Nonoperator landlords accounted for the remaining 42 percent of land in farms. Ninety-five percent of non-farm landlords were individuals/families or partnerships. Of these unincorporated landlords, 55 percent were at least 65 years old. Many non-farm landlords have a historic connection to farming. Among the people who have exited farming or inherited farmland since the number of farms peaked during the Great Depression, a number have retained ownership of some or all their land (Hoppe et al., 1995).

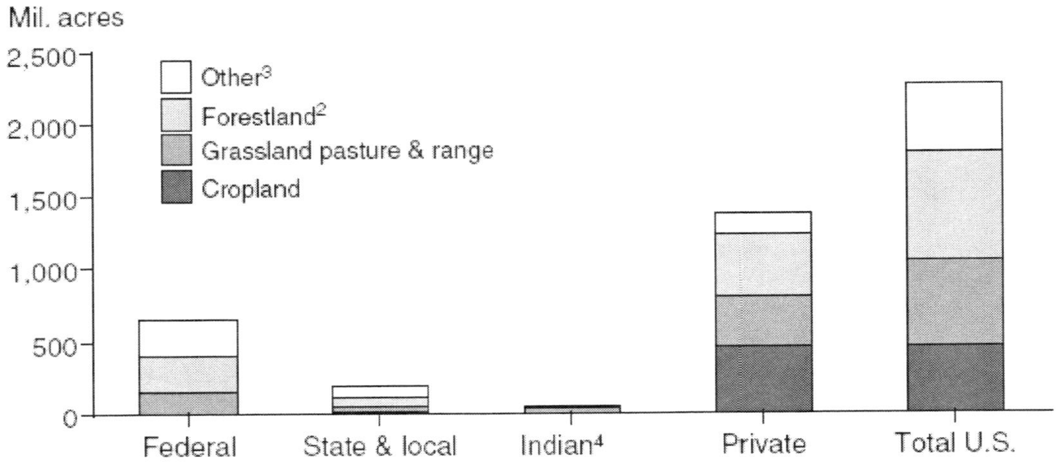

[1] All 50 States.

[2] Includes forestland in parks and other special uses.

[3] Includes urban land, highways, and other special or miscellaneous uses. Excludes an estimated 105 million acres in special uses that have forest cover, and therefore are included with forestland.

[4] Managed in trust by the Bureau of Indian Affairs.

Source: USDA, ERS, based on Major Land Use estimates (Lubowski et al., 2006).

Figure 7. Major uses of U.S. land by ownership, 19971.

FARM NUMBERS, FARM TYPES, AND CONSERVATION PROGRAMS

The number of farms has declined dramatically since its peak of 6.8 million in 1935, with most of the decline occurring during the 1940s, 1950s, and 1960s (Figure 8).

The decline in farm numbers has leveled off since the 1970s. By 2002, 2.1 million farms remained. The remaining farms have a much larger average acreage, but averages mask differences among farms. Today's farms range from very small retirement and residential farms to industrialized operations with sales in the millions. Part of this diversity stems from the very low sales threshold ($1,000) necessary for an operation to qualify as a farm for statistical purposes.

One way to address the diversity of farms is to categorize them into more homogeneous groups. The farm typology developed by ERS identifies five groups of small family farms (sales less than $250,000): limited-resource, retirement, residential/lifestyle, farming-occupation/low-sales, and farming-occupation/high-sales (see box, "Farm Typology Definitions"). The typology also includes large family farms, very large family farms, and

The farm typology focuses on the "family farm," or any farm organized as a sole proprietorship, partnership, or family corporation. Family farms exclude farms organized as nonfamily corporations or cooperatives, as well as farms with hired managers.

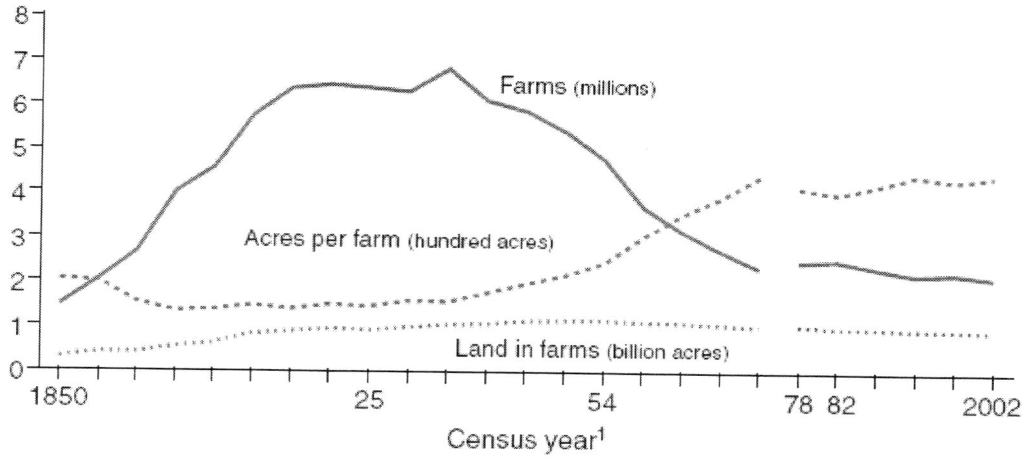

[1] Census years are shown for 10-year intervals from 1850 to 1920, with 4 or 5-year intervals starting in 1925. The break in the lines after 1974 reflects the introduction of an adjustment to estimates of the farm count and land in farms. Beginning in 1978, the data are adjusted to compensate for undercoverage by the census of agriculture.

Source: USDA, ERS, based on census of agriculture data.

Figure 8. Farms, land in farms, and average acres per farm, 1850-2002.

Farm Typology Definitions

The farm typology focuses on the "family farm," or any farm organized as a sole proprietorship, partnership, or family corporation. Family farms exclude farms organized as nonfamily corporations or cooperatives, as well as farms with hired managers.

Small family farms

Limited-resource farms. Small farms with sales less than $100,000 and low operator household income (defined as less than the poverty level for a family of four in the current and previous years or less than half the county median household income both years.)

Retirement farms. Small farms whose operators report they are retired.[1]

Residential/lifestyle farms. Small farms whose operators report a major occupation other than farming.[1]

Farming-occupation farms. Small family farms whose operators report farming as their major occupation.[1]

- **Low-sales farms.** Sales less than $100,000.
- **High-sales farms.** Sales between $100,000 and $249,999.

Other family farms

Large family farms. Sales between $250,000 and $499,999.

Very large family farms. Sales of $500,000 or more.

Nonfamily farms

Nonfamily farms. Farms organized as nonfamily corporations or cooperatives, as well as farms operated by hired managers.

[1] Excludes limited-resource farms whose operators report this occupation.

For more information about the farm typology, see the *2004 Family Farm Report* (Banker and MacDonald, 2005).

Size Variation Among Typology Groups

Small family farms dominate the farm count, making up 91 percent of all U.S. farms in 2003 (Table 5).

In addition, very small farms (sales less than $10,000) make up more than half of all farms. Very small farms account for a particularly large share of farms in the limited-resource (72 percent), retirement (76 percent), and residential/lifestyle (76 percent) groups. Production, however, is concentrated among larger farms; small farms account for only 27 percent of the total value of production.

The smallness of most farms has implications for conservation and the environment. An ERS study found that smaller corn farms are less likely to use conservation tillage than are larger farms (Soule et al., 1999 and 2000). The practice is more practical for larger farms because they have more acres over which to spread the cost of new or retrofitted equipment necessary to adopt conservation tillage. Small farms whose operators are retired or farm part-time are also less likely to adopt conservation tillage, possibly because of hesitancy to change familiar production practices. Small farms, however, participate widely in the Conservation Reserve Program (CRP) and the Wetlands Reserve Program (WRP). (For more information about conservation tillage, see Chapter 13, "Soil Management and Conservation".)

Table 5. Selected farm structural characteristics, by the farm typology, 2003

Farm typology group	Farms	Value of production	Sales less than $10,000	Tenure: Full owner	Tenure: Part owner	Tenant[1]
Small family farms:	Pct. of U.S. total			Percent of group		
Limited-resource	11.1	1.4	71.8	68.8	24.3	*6.9
Retirement	14.6	1.5	75.6	79.0	19.4	1.6
Retirement/lifestyle	42.1	5.2	75.8	70.6	25.5	3.9
Farming-occupation:						
Low-sales	17.2	6.6	37.0	54.9	36.5	8.6
High-sales	6.4	12.3	na	19.1	68.2	12.7
Large family farms	4.0	14.4	na	20.9	66.4	12.6
Very large family farms	3.1	44.7	na	24.1	58.7	17.2
Nonfamily farms	1.7	13.7	31.9	65.5	23.7	10.8
All farms	100.0	100.0	57.7	62.1	31.7	6.1

na = Not applicable.

* = Standard error is between 25 and 50 percent of the estimate.

[1]Farms that rent all the land that they operate. Also includes farms owning less than 1 percent of the land they operate.

Source: USDA, ERS, 2003 Agricultural Resource Management Survey, Phase III.

Distribution of Conservation Program Payments by Type of Farm

High-sales small farms, large family farms, and very large family farms received a disproportionate share of commodity program payments relative to their small share of farms in 2003 (Table 6).

Table 6. Share of government payments and related items, by the farm typology, 2003

Farm typology group	Government payments: Commodity programs[1]	Government payments: Conservation programs[2]	Harvested acres of program crops[3]	Land enrolled in CRP and WRP
	Percent of U.S. total			
Small family farms:				
Limited-resource	2.1	6.6	2.4	5.7
Retirement	2.1	19.9	1.8	22.5
Residential/lifestyle	6.3	26.4	6.4	25.7
Farming-occupation:				
Low-sales	10.1	17.6	9.9	18.7
High-sales	22.3	9.6	23.3	9.5
Large family farms	22.6	8.5	23.9	8.5
Very large family farms	31.8	7.2	29.8	5.4
Nonfamily farms	2.7	4.2	2.6	4.0
All farms	100.0	100.0	100.0	100.0

[1]Direct payments, countercyclical payments, loan deficiency payments, marketing loan gains, net value of commodity certificates, peanut quota buyout, milk income loss contract payments, etc.

[2]Payments from the Conservation Reserve Program (CRP), the Wetlands Reserve Program (WRP), and the Environmental Quality Incentives Program (EQIP).

[3]Food and feed grains, soybeans, other oilseeds, sugar beets, and sugar cane.

Source: USDA, ERS, 2003 Agricultural Resource Management Survey, Phase III.

These farms harvest most of the land planted to program commodities and therefore receive three-quarters of commodity program payments. However, CRP and WRP—the two major conservation programs—are targeted at particular types of land, not commodities. Since small farms own 70 percent of the land held by farms, they play a large role in natural resource and environmental policy. (For more information about CRP and WRP, see Chapter 22, "Land Retirement Programs".)

Retirement, residential/lifestyle, and low-sales farms account for nearly two-thirds of conservation payments and a similar share of the land farmers enrolled in the CRP and WRP. Participating farmers in each of the three groups tend to enroll large shares of their land in these programs: 46 percent of the land operated on retirement farms, 28 percent on residential/lifestyle farms, and 23 percent on low-sales farms. In contrast, enrollment ranges from 5 to 9 percent for participating high-sales, large, and very large farms.

Because their main job is off-farm, residential/lifestyle operators are limited in the amount of time they can spend farming. As a result, residential/lifestyle farmers find CRP and WRP attractive, since these programs require little time. Given their life-cycle position, many retired farmers have land available to put into conservation uses. The same forces may also be acting on low-sales operators, who average 57 years of age and may be scaling down their operations.

If an off-farm job and advanced age are major determinants of land going into conservation uses, it may be relatively easy to get smaller farms to enroll land in the programs. Getting larger farms to enroll more of their land might require higher payments, if the opportunity cost of idling their land is higher.

LAND TENURE

Farm operators leased 38 percent of their total farmland in 2002, down from 40 percent in 1997 and 43 percent in 1992, according to the census of agriculture. This decline may reflect increasing rental costs as parcels of land become smaller. Parcels of farmland available to rent tend to become subdivided with time due to division among heirs (Raup, 2003). Smaller parcels increase transaction costs to operators assembling land to expand their operations. Still, rented land as a share of total farmland is higher than the 35-percent rate that prevailed in the 1950s and 1960s.

About 38 percent of all farms rented land in 2003, 32 percent as part owners and 6 percent as tenants (Table 5). Land leasing has changed from a way for beginning farmers to enter agriculture to a way for established farmers to access additional land. Renting allows farms to expand without the debt and commitment of capital associated with ownership (Reimund and Gale, 1992). In fact, about 17 percent of very large family farms are tenants, a larger percentage than in any other group.

Conventional wisdom holds that farmland owners have a long-term interest in their land and thus are more likely than renters to adopt conservation practices. Soule and others (1999 and 2000) found this to be true among corn farmers, at least in the adoption of conservation practices that provide only long-term benefits, such as grassed waterways and strip cropping.

The situation was different for conservation tillage, which can increase profits in the short run by maintaining or increasing yields while reducing machinery, fuel, and labor costs (Magleby, 2003). Cash-renters are less likely than owner-operators to use conservation tillage, but share-renters appear to act like owner-operators in adopting conservation tillage. Share-renters may have an incentive to adopt conservation tillage, if the landlord bears some of the costs that may increase under conservation tillage, such as herbicide expenditures. Share-landlords are also more likely to be involved in management decisions than cash-landlords, which may make share-renters act more like owners.

CONCENTRATION OF PRODUCTION

Concentration of agricultural production on fewer farms and fewer acres has grown since the beginning of the 20th century. In 1900, half of farm sales came from approximately 17 percent of farms and 43 percent of the land in farms (Figure 9).

By 2002, half of farm sales came from 2 percent of U.S. farms and 11 percent of the land in farms. This reflects both a growing diversity in farm size and an increasing number of very large farms.

The concentration of agricultural production raises concerns about potential harm to the environment, especially from livestock operations. Data from the census of agriculture show that the number of U.S. farms selling hogs decreased by 94 percent between 1959 and 2002, while hog sales more than doubled. Similar trends have occurred among farms selling dairy products, cattle, and broilers. As livestock producers expand, they are more likely to buy feed grown elsewhere, reducing the amount of land they have available for manure application, the predominant method of disposal (Ribaudo et al. 2003).

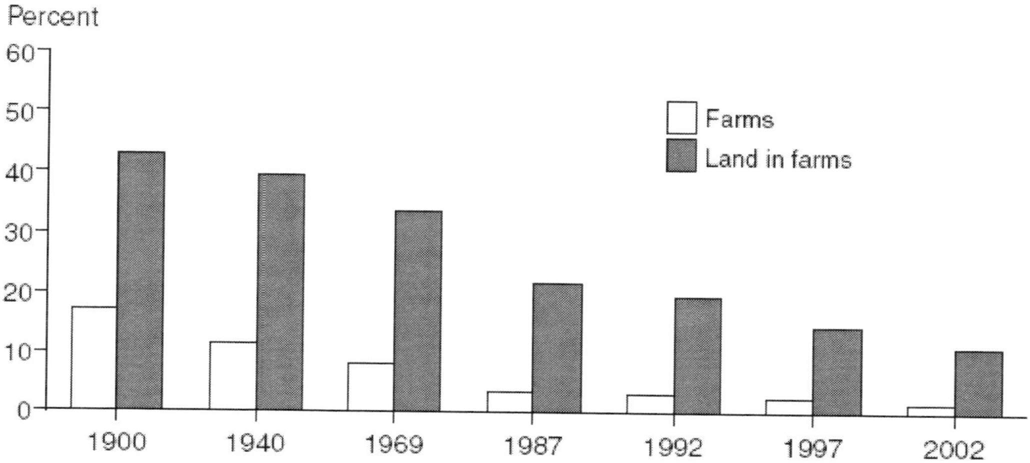

Note: The share of sales in 1900, 1949, and 1969 was calculated by summing sales by sales class from census publications and totaled slightly more than 50 percent. The share of sales in 1987, 1992, 1997, and 2002 was calculated from farm-level data and therefore totaled exactly 50 percent.

Source: USDA, ERS, compiled from census of agriculture data and Peterson and Brooks (1993).

Figure 9. Share of U.S. farms and land in farms producing half of the Nation's agricultural sales.

More livestock production on fewer farms may not pose a problem if farms with livestock have enough land to absorb the manure produced. In fact, most farms currently have adequate land to safely use the manure that their livestock produce. Many livestock producers, however, do not apply manure to all their land (Ribaudo, 2003). Manure is expensive to haul, so many producers spread more manure than crops need on the fields nearest the livestock facility. In addition, adequate farmland for manure disposal may not exist in some areas with large concentrations of livestock. For example, there are 68 counties where nitrogen in manure from confined livestock and poultry farms is estimated to exceed the county's nitrogen needs. Excess phosphorus is even more common, occurring in 152 counties. (For more information, see Chapter 16, "Animal Agriculture and the Environment".)

References

[1] Banker, David E., and James M. MacDonald (2005). *Structural and Financial Characteristics of U.S. Farms: 2004 Family Farm Report.* AIB-797. U.S. Dept. Agr., Econ. Res. Serv. March.

[2] Hoppe, Bob, Bob Green, and Gene Wunderlich (1995). "Farmland Rentals: Central to Farming," *Agricultural Outlook*. AO-220. U.S. Dept. Agr., Econ. Res. Serv. July.

[3] Lubowski, Ruben N., Marlow Vesterby, Shawn Bucholtz, Alba Baez, and Michael J. Roberts (2006). *Major Uses of Land in the United States, 2002*. EIB-14. U.S. Dept. Agr., Econ Res. Serv., May.

[4] Magleby, Richard (2003). "Chapter 4 Soil Management and Conservation," *Agricultural Resource and Economic Indicators, 2003*. Ralph Heimlich (ed.). AH-722. U.S. Dept. Agr., Econ. Res. Serv. Feb.

[5] Peterson, R. Neal, and Nora L. Brooks (1993). *The Changing Concentration of U.S. Agricultural Production During the 20th Century: 14th Annual Report to the Congress on the Status of the Family Farm*. AIB-671. U.S. Dept. Agr., Econ. Res. Serv. July.

[6] Raup, Philip M. (2003). "Disaggregating Farmland Markets," *Government Policy and Farmland Markets: The Maintenance of Farmer Wealth*. Charles B. Moss and Andrew Schmitz (eds.). Ames, IA: Iowa State Press.

[7] Reimund, Donn A., and Fred Gale (1992). *Structural Change in the U.S. Farm Sector, 1974-87: 13th Annual Family Farm Report to Congress*. AIB-647. U.S. Dept. Agr., Econ. Res. Serv. May.

[8] Ribaudo, Marc (2003). "Managing Manure: New Clean Water Act Regulations Create Imperative for Livestock Producers," *Amber Waves*. Vol. 1, No. 1, Feb.

PART 2: WATER AND WETLAND RESOURCES

In: Agricultural Resources and Environmental Indicators ISBN 978-1-60021-467-7
Editor: Keith Wiebe and Noel Gollehon, pp. 29-37

Chapter 4

IRRIGATION RESOURCES AND WATER COSTS

Noel Gollehon and William Quinby

Irrigated agriculture remains the dominant use of freshwater in the United States, although its share of use is declining. Irrigated cropland area has expanded over 40 percent since 1969, while water application rates have declined about 20 percent. The total quantity of irrigation water applied increased about 10 percent since 1969. Nationally, the average variable cost of supplying water for irrigation was about $50 per acre in 2003; however, that amount does not reflect the full value of water.

INTRODUCTION

The United States, as a whole, has abundant freshwater supplies. Annual renewable supplies in surface streams and aquifers total roughly 1,500 million acre-feet per year (maf/yr). (See the "Irrigation and Water Use"Briefing Room on the ERS website) for definitions. Of total renewable supplies, only one-quarter is withdrawn for use in homes, farms, and industry, and just 7 percent is actually used, i.e., lost to the immediate water environment (Moody, 1993). Roughly 90 percent of total water use nationwide comes from renewable surface- and ground-water supplies. The remainder comes from depletion of stored ground water (Foxworthy and Moody, 1986).

An abundance of water in the aggregate belies increasingly limited water supplies in many areas, reflecting the uneven distribution of the Nation's water resources. In the arid West, more than half of the renewable water supplies are consumed under normal precipitation conditions. In drought years, water use often exceeds renewable flow through the increased use of water stored in aquifers and reservoirs. While droughts exacerbate supply scarcity, water demands continue to expand with resulting reallocations among uses. Urban growth, for example, has greatly expanded municipal water demands in arid areas of the Southwest and far West. At the same time, demand for instream (nonconsumptive) water flows for recreation, riparian habitat, and other environmental purposes has heightened competition for available water supplies in all but the wettest years. While future water needs for instream uses are difficult to quantify, the potential demands on existing water supplies are large and geographically diverse.

Historically, increased water demands were met by expanding available water supplies. Dam construction, groundwater pumping, and interbasin conveyance provided the water to meet growing urban and agricultural needs. However, future opportunities for large-scale expansion of seasonally reliable water supplies are limited due to lack of suitable project sites, limited funding, and increased public concern for environmental consequences. Future water demands will increasingly be met through reallocation of existing supplies. Since agriculture is the largest freshwater user, reallocation will likely reduce supplies for agriculture (National Research Council, 1996). Changes in agricultural water availability may have significant impacts on irrigation-dependent crops in some locations, with implications for local agricultural industries and rural communities.

Agricultural Water Withdrawals

Freshwater withdrawals—the quantity of water diverted from surface and ground-water sources—totaled 387 million acre-feet (maf) in 2000 (Figure 10). Agriculture (159 maf) and thermoelectric power generation (152 maf) dominate withdrawals, with domestic and commercial water supplies, industry, and mining withdrawing a combined 75 maf (Hutson et al., 2004).

Agricultural withdrawals as a share of U.S. freshwater withdrawals declined from 46 percent in 1960 to 41 percent in 2000[1]. Thermoelectric power generation increased its share from 32 to 39 percent over the same period. Water withdrawals are not the only measure of water use. Consumptive use—the water not returned to the immediate water environment—is much greater for agriculture than any other sector, both in total and as a share of water withdrawn. Estimates available from 1960 through 1995 show that agriculture accounts for over 80 percent of the Nation's consumptive use (Figure 10), because a high share of applied irrigation water is used by plants for evapotranspiration, with little returning to surface or ground water. (Water diverted for cooling thermoelectric plants tends to be used as a thermal sink, with much of it returned to rivers and streams.) Greater irrigation withdrawals do not necessarily translate into greater consumptive use per irrigated acre. The difference between withdrawals and consumptive use highlights the importance of losses, runoff, and return flows.

Most agricultural water withdrawals occur in the arid Western States where irrigated production is concentrated. In 2000, about 85 percent of total agricultural withdrawals occurred in a 19-State area encompassing the Plains, Mountain, and Pacific regions (Table 7). In the Mountain region, over 90 percent of the water withdrawn is used by agriculture, almost all (96 percent) for irrigation. Nationally, irrigation is the dominant agricultural water use, but water withdrawn for livestock and aquaculture production (including fish hatcheries) accounts for almost 20 percent of withdrawals in the North-Central and Eastern States. Even in these more humid States, irrigation is the dominant agricultural water use.

[1] The irrigation component of the withdrawal estimates by Hutson et al. are primarily agricultural (cropland and pastureland), but also include recreational area irrigation (parks and golf courses).

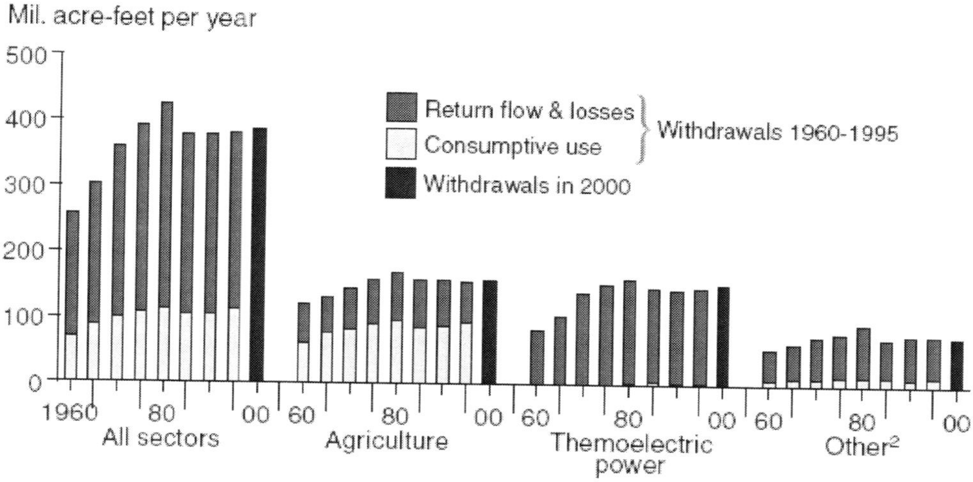

[1]Data limitations do not allow estimation of consumptive use in 2000.

[2]Includes public supplies, domestic supplies and industry, except power generation.

Source: USDA, ERS, based on Hutson et al., 2004.

Figure 10. Water withdrawals in 2000 and withdrawals with consumptive use estimates, 1960-1995[1].

Surface water accounted for 59 percent of total irrigation withdrawals in 2000, with ground water supplying the remainder. Ground water is a growing source of agricultural water supplies, increasing from 37 to 41 percent of total withdrawals since 1960. Ground water supplied most of the irrigation water in the eastern 37 States, the area experiencing the most irrigation growth in the past decade. In the Pacific and Mountain regions, surface-water supplies are still the dominant water source (Table 7).

Table 7. Agricultural water withdrawals, by region and total U.S., 2000

Region	Number of States	Agricultural water withdrawals		Components of agricultural withdrawals		Source of agricultural withdrawals	
		Share of total withdrawals	Quantity	Irrigation	Livestock and aquaculture	Ground water	Surface water
		Percent	1,000 acre-feet per year	Percent			
Pacific	5	80	45,879	98	2	34	66
Mountain	8	91	64,209	96	4	20	80
Plains	6	49	25,901	97	3	80	20
South	7	30	19,054	95	5	73	27
North-Central & East	24	3	4,409	81	19	72	28
U.S. total[1]	50	41	159,558	96	4	41	59

[1]Excludes water withdrawals in the U.S. Virgin Islands, Puerto Rico, and the District of Columbia.

Source: USDA, ERS, based on Hutson et.al., 2004.

Environmental harm can occur whenever water is withdrawn for agriculture (or any other extractive use). Surface-water withdrawals include either the gravity diversions of rivers and streams or the pumping of water from lakes, rivers, or streams, which can reduce (or totally dry up) streamflow and impair species habitat and wetlands. Ground water is withdrawn with pumps from wells drilled into underground water-bearing strata. When withdrawals exceed natural rates of aquifer recharge, the extraction of ground water can cause land subsidence, reduce total water reserves, and reduce base stream-flow, thereby triggering surface-water shortages.

IRRIGATED LAND AND ASSOCIATED WATER

In 2002, U.S. irrigated farmland occupied 55.3 million acres, down 1 million acres from 1997 (Table 8). Despite this recent decline, irrigated farmland has increased at an average rate of a half million acres per year over the last three decades, continuing a century-long trend (Figure 11).

Table 8. Irrigated land in farms, by region and crop, selected years 1969-2002

Region or	1969[1]		1997[2]		2002[2]	
crop	1,000 acres	Percent	1,000 acres	Percent	1,000 acres	Percent
United States[3]	39,100	100	56,289	100	55,311	100
Region						
Eastern regions[4]	4,200	11	12,308	22	13,797	25
Northern Plains	4,600	12	10,312	18	10,907	20
Southern Plains	7,400	19	6,273	11	5,592	10
Mountain	12,800	33	13,603	24	12,079	22
Pacific Coast	10,000	26	13,713	24	13,372	24
Crop						
Corn for grain	3,200	8	10,816	19	9,710	18
Other grains	9,200	24	9,245	16	7,703	14
Soybeans	700	2	4,238	8	5,460	10
Cotton	3,100	8	5,152	9	4,802	9
Alfalfa hay	5,000	13	6,087	11	6,809	12
Vegetables and orchards	3,900	10	6,722	12	6,734	12
Other lands in farms[5]	14,000	36	14,030	25	14,093	25

[1]Census of Agriculture.

[2]Census of Agriculture, adjusted for non-response.

[3]Includes Alaska and Hawaii.

[4]Northeast, Appalachian, Southeast, Lake States, and Corn Belt.

[5]Other uses with more than 500,000 irrigated acres include corn silage, other hay, dry beans, potatoes, sugar beets, nursery crops, cropland pasture, and other pasture.

Source: USDA, Census of Agriculture, selected years.

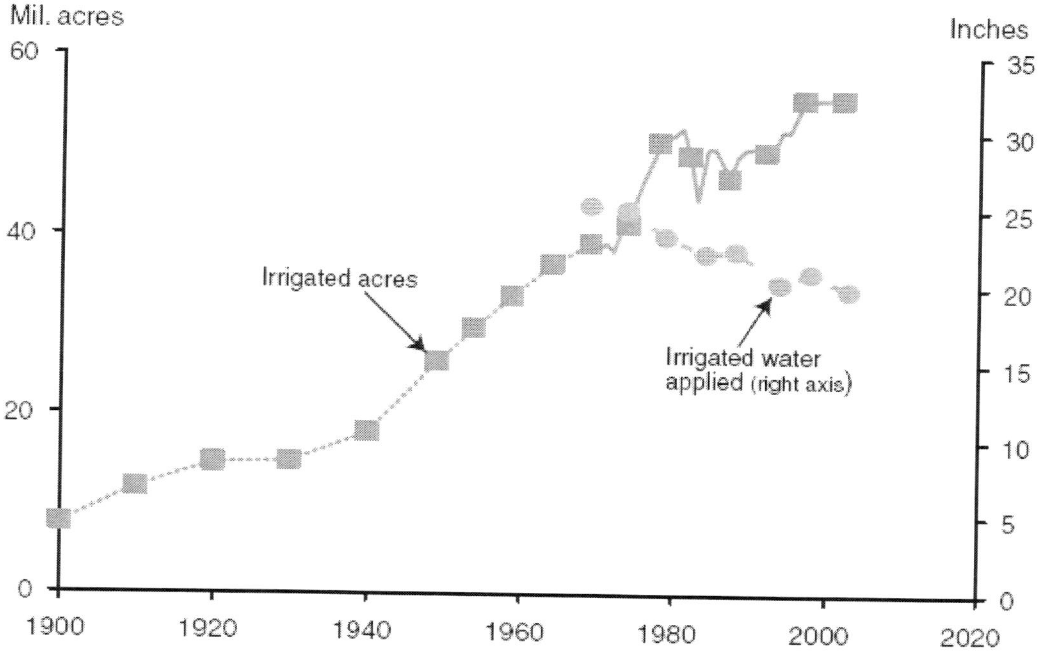

Source: USDA, Census of Agriculture and Farm and Ranch Irrigation Surveys, various years. Variation between Census of Agriculture years from 1969 to 2002 was based on ERS estimates.

Figure 11. Trends in acres irrigated from 1900 to 2002 and water applied from 1969 to 2003.

Substantial variation within the trend can largely be explained by year-to-year changes in four factors: farm program requirements, crop prices, water supplies in the West, and weather influences on the need for supplementary irrigation in humid areas.

In recent years, national irrigated area has plateaued at about 55 million acres as continuing growth in eastern States has been offset by declines in western irrigation. Since 1988, western irrigated area has been affected by two extended droughts that led to water supply problems, especially in the Southwest. In general, there is an increasing reliance on irrigation in the humid East, and a northward redistribution of irrigation in the West (Figure 12). In recent decades, large concentrations of irrigation have emerged in humid areas—Florida, Georgia, and especially in the Mississippi Delta, primarily Arkansas and Mississippi.

Averaged over all States and crops, the average depth of water applied has declined by one-fifth (5.4 inches per-acre) since 1969, to annual application levels of less than 20 inches in 2003 (Figure 11). Agriculture has adopted more water-conserving practices and shifted irrigated production of some commodities to more humid and cooler areas, requiring less supplementary water. Irrigation application can vary from less than 6 inches per acre (sorghum in the North-Central States) to more than 4.5 feet per acre (orchards in the Mountain States). (Water use rates in 2003 were affected by extended drought in the West, especially the Southwest. Surface-water allocations dipped below 50 percent of normal levels in some areas.)

Changes in total water applied to irrigated lands reflect efficiency gains per acre, shifts in crop locations, and changes in acres irrigated. Per-acre declines in application rates (see Chapter 17, "Irrigation Water Management") have partially offset the increase in irrigated acreage since 1969. Over 1969-2003, irrigated acreage increased by over 40 percent while total water applied increased by only 11 percent.

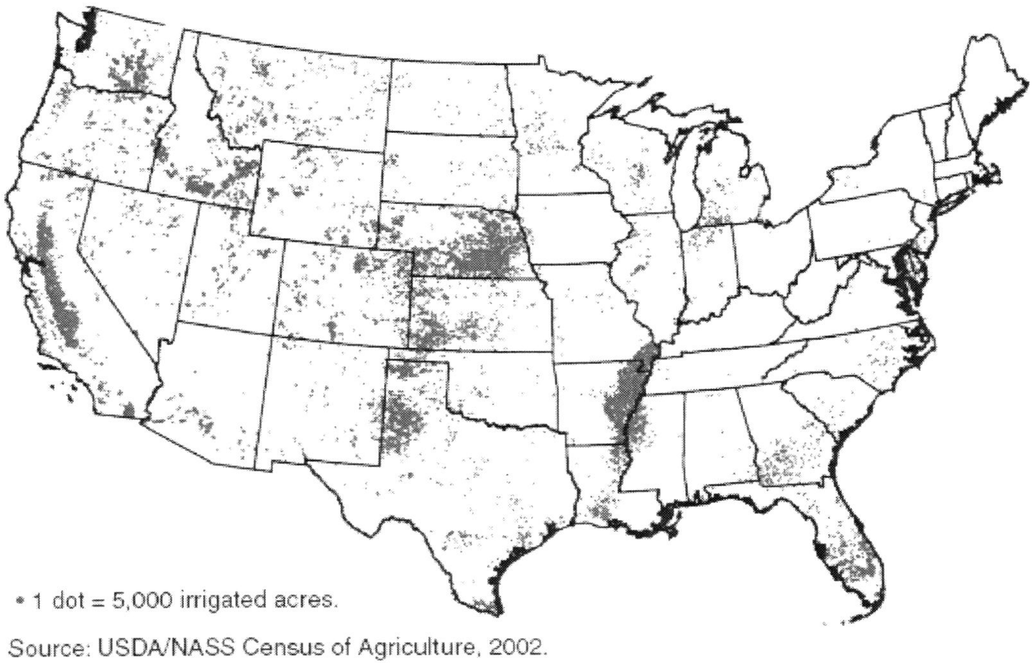

Source: USDA/NASS Census of Agriculture, 2002.

Figure 12. Distribution of irrigated land in farms, 2002.

Irrigation Water Prices and Costs

Prices paid for irrigation water are of considerable policy interest due to their importance as a cost of production and their impact on water demand. Increasingly, adjusting the water "price" is viewed as a mechanism to improve the economic efficiency of water use. However, water price adjustments to achieve socially desired outcomes can be difficult because prices paid for water are rarely set in a market and generally do not convey signals about water's scarcity. States generally administer water resources and grant (not auction) rights of use to individuals without charge, except for minor administrative fees. As a result, expenditures for irrigation water usually reflect water's access and delivery costs alone—thus, costs to irrigators usually do not reflect the full social cost of water use. (By contrast, those without an existing State-allocated water right—whether an irrigator, municipality, industry, or environmental group—that purchase annual water allocations or permanent water rights from existing users pay prices that more closely reflect the scarcity value of the resource.)

Costs of supplying irrigation water vary widely, reflecting different combinations of water sources, suppliers, distribution systems, and other factors such as field proximity to water, topography, aquifer conditions, and energy source. To generalize, ground water is usually pumped on-farm with higher energy expenses than surface water, which is often supplied from off-farm sources through extensive storage and canal systems. We use data from the Farm and Ranch Irrigation Survey (USDA, 2004b) to examine the cost determinants for ground- and surface-water sources.[2]

Table 9. Costs of irrigation water by source and category, 2003

Cost category	Acres incurring the cost Million	Percent	State-level cost range Dollars per acre	National average cost Dollars per acre	Total national cost $ million
Energy expenses for pumping ground water	32.34	61.5	7- 176	39.50	1,277.54
Energy expenses for lifting or pressurizing surface water	10.56	20.1	10 - 82	26.39	278.72
Water purchased from off-farm sources	13.87	26.4	5 - 86	41.73	578.75
Maintenance/repair expenses	40.01	76.1	4 - 80	12.29	491.77
Total variable costs					2,622.37
Average variable cost (including acres with no cost)				49.87	
Capital investment expenses[1] incurred in 2003	26.67	50.7	16 - 187	42.18	1,125.13

[1]Over $13,000 per farm, distributed based on average farm size to compute per-acre expenses.
Source: USDA, ERS, based on the 2003 Farm and Ranch Irrigation Survey, USDA (2004b).

POLICY ISSUES

Several types of organizations serve as "off-farm suppliers" of water to irrigators, but most are nonprofits that provide dependable water service at low cost. Some such organizations have developed extensive regional water storage and conveyance facilities, while others serve as a local water retailer, transferring water from a wholesaler (such as the Bureau of Reclamation) to water users. Water pricing by these organizations is often based on acreage served rather than water delivered, since administrative costs are lower with acreage-based charges. With this pricing system, producers have little financial incentive to conserve water since charges are assessed independently of how much water allotment is used.

[2] Acres irrigated reported in the 2003 Farm and Ranch Irrigation Survey (FRIS) exclude certain types of irrigated farms accounting for about 10 percent of the irrigated land reported in the 2002 Census of Agriculture. FRIS is the sole data source reporting both cost information and acres irrigated by water source.

The Bureau of Reclamation (Reclamation), U.S. Department of the Interior, is the primary Federal agency involved in developing and managing water supply projects for irrigation purposes. Reclamation serves as a water "wholesaler" for about 25 percent of the West's irrigated acres—collecting, storing, and conveying water to local entities that, in turn, serve irrigators. From 1902 through 1994, the Reclamation program constructed 133 projects that provide irrigation water, costing $21.8 billion. Irrigation is scheduled to pay less than half of its allocated share of construction costs, with most of the cost subsidized by hydropower revenue (General Accounting Office, 1996). New demands on water for urban growth and environmental restoration in areas with Reclamation projects have focused attention on issues such as the recovery of water-supply subsidies, improved economic efficiency, and increased conservation through water pricing.

Increasing water demands for urban and environmental purposes have prompted discussions on how to more accurately reflect the opportunity costs of water in prices paid by irrigators. Several options exist for States (and in some cases Reclamation) to modify price or quantity allocations to more accurately reflect the scarcity value of water and to improve social benefits.

Voluntary water markets are one prominent strategy to meet new water needs. However, current markets have transactions totaling only 1 to 2 percent of irrigation withdrawals, with volumes concentrated in a few States (Howitt and Hansen, 2005). Markets are most active in areas where there are fewer barriers (defined property rights, institutional flexibility, and developed physical infrastructure), or demand is such that participants are willing to pay significant transaction costs. The most prevalent type of exchange, with nearly 90 percent of the volume, is water leases (especially annual transfers), with permanent transfer of water rights and option markets the remainder.

Irrigated agriculture is likely to remain important, both in terms of the value of agricultural production and demand on land and water resources (National Research Council, 1996). However, changes in the irrigation sector are anticipated in response to increasing water demands for urban and environmental uses, as well as evolving institutions governing farm programs and water allocations. Water diversions for agricultural production will likely continue to decline, with at least some portion shifted to satisfy alternative goals.

REFERENCES

[1] Foxworthy, B.L., and D.W. Moody (1986). "Water-Availability Issues: National Perspective on Surface-Water Resources," *National Water Summary 1985—Hydrologic Events and Surface-Water Resources*. U.S. Geological Survey, U.S. Department of the Interior. Water Supply Paper 2300, pp. 51-68.

[2] Howitt, R., and K. Hansen (2005). "The Evolving Western Water Markets," *Choices*. 20(1):59-63.

[3] Hutson, S.S., N.L. Barber, J.F. Kenny, K.S. Linsey, D.S. Lumia, and M.A. Maupin (2004). *Estimated Use of Water in the United States in 2000*. U.S. Geological Survey, U.S. Department of the Interior. Circular 1268.

[4] Moody, D.W. (1993). "Water: Fresh-Water Resources of the United States," *Research & Exploration*. Water Issue, National Geographic Society, Washington, DC, pp. 80-85.

[5] National Research Council Water Science and Technology Board (1996). *A New Era for Irrigation.* Washington, DC.

[6] U.S. Department of Agriculture, Economic Research Service (1997). *Agricultural Resources and Environmental Indicators, 1996-97.* AH-712

[7] U.S. Department of Agriculture, National Agricultural Statistics Service (2004a). *2002 Census of Agriculture.* Volume 1, Geographic Area Series, Part 51, AC02-A-51.

[8] U.S. Department of Agriculture, National Agricultural Statistics Service (2004b). *Farm and Ranch Irrigation Survey (2003).* Volume 3, Part 1, Special Studies of 2002 Census of Agriculture, AC-02-SS-1.

[9] U.S. Department of Agriculture, National Agricultural Statistics Service (1999a). *1997 Census of Agriculture.* Volume 1, Geographic Area Series, Part 51, AC97-A-51.

[10] U.S. Department of Agriculture, National Agricultural Statistics Service (1999b). *Farm and Ranch Irrigation Survey (1998).* Volume 3, Part 1, Special Studies of 1997 Census of Agriculture, AC97-SP-1.

[11] U.S. General Accounting Office (1996). *Bureau of Reclamation: Information on Allocation and Repayment of Costs of Constructing Water Projects.* GAO/RCED-96-109. Washington, DC.

In: Agricultural Resources and Environmental Indicators ISBN 978-1-60021-467-7
Editor: Keith Wiebe and Noel Gollehon, pp. 39-47

Chapter 5

WATER QUALITY: IMPACTS OF AGRICULTURE

Marc Ribaudo and Robert Johansson

Agricultural production releases residuals, like sediment and pesticides, that may degrade the quality of water resources and impose costs on water users. Agriculture is the leading source of impairments in the Nation's rivers and lakes and a major source of impairments to estuaries. However, the extent and magnitude of this degradation is difficult to assess because of its nonpoint nature.

INTRODUCTION

The production practices and inputs used by agriculture can result in a number of pollutants entering water resources, including **sediment, nutrients, pathogens, pesticides, and salts**. Farmers, when making production decisions, often do not consider offsite impacts associated with runoff or leaching. Documenting the links between agriculture and water quality can help policymakers provide appropriate incentives to farmers for controlling pollution that originates on farms.

Agriculture is widely believed to have significant impacts on water quality. While no comprehensive national study of agriculture and water quality has been conducted, the magnitude of the impacts can be inferred from several water quality assessments. A general assessment of water quality is provided by EPA's 2000 Water Quality Inventory. Based on State assessments of 19 percent of river and stream miles, 43 percent of lake acres, and 36 percent of estuarine square miles, EPA concluded that agriculture is the leading source of pollution in 48 percent of river miles, 41 percent of lake acres (excluding the Great Lakes), and 18 percent of estuarine waters found to be water-quality impaired, in that they do not support designated uses. This makes agriculture the leading source of impairment in the Nation's rivers and lakes, and a major source of impairment in estuaries. Agriculture's contribution has remained relatively unchanged over the past decade.

The significance of water pollutants commonly produced by agriculture is suggested by information on impaired waters provided by States, tribes, and territories to EPA in accordance with Section 303(d) of the Clean Water Act. These are waters that do not meet water quality standards, and cannot meet those standards through point-source controls alone. The most recent information (2005) indicates that 25,823 bodies of water (stream reaches or lakes) are impaired nationwide. Pathogens, sediment, and nutrients are among the top sources of impairment, and agriculture is a major source of these pollutants in many areas.

A U.S. Geological Survey (USGS) study of agricultural land in watersheds with poor water quality estimated that 71 percent of U.S. cropland (nearly 300 million acres) is located in watersheds where the concentration of at least one of four common surface-water contaminants (nitrate, phosphorus, fecal coliform bacteria, and suspended sediment) exceeded generally accepted instream criteria for supporting water-based recreation activities (Smith, Schwarz, and Alexander, 1994). Another USGS study found that structural changes in animal agriculture between 1982 and 1997 put upward pressure on stream concentrations of fecal coliform bacteria in many areas of the Great Plains, Ozarks, and Carolinas (Smith et al., 2005).

MAJOR AGRICULTURAL POLLUTANTS

Sediment is the largest contaminant of surface water by weight and volume (Koltun et al., 1997) and is identified by States as the second leading pollution problem in rivers and streams and the third leading problem in lakes (USEPA, 2002). Sediment in surface water is largely a result of soil erosion (see Chapter 13, "Soil Management and Conservation"), which is influenced by soil properties and the production practices farmers choose. Sediment buildup reduces the useful life of reservoirs. Sediment can clog roadside ditches and irrigation canals, block navigation channels, and increase dredging costs. By raising streambeds and burying streamside wetlands, sediment increases the probability and severity of floods. Suspended sediment can increase the cost of water treatment for municipal and industrial water uses. Sediment can also destroy or degrade aquatic wildlife habitat, reducing diversity and damaging commercial and recreational fisheries.

Regions with the greatest potential to discharge sediment from cropland to surface waters include parts of the Heartland, Mississippi Portal, and Prairie Gateway (see the ERS website for a description of Farm Resource Regions for a description) (Figure 13).

Nitrogen and **phosphorus** are important crop nutrients, and farmers apply large amounts to cropland each year. They can enter water resources through runoff and leaching. The major concern for surface-water quality is the promotion of algae growth (known as eutrophication), which can result in decreased oxygen levels, fish kills, clogged pipelines, and reduced recreational opportunities. USGS has found that high concentrations of nitrogen in agricultural streams are correlated with nitrogen inputs from fertilizers and manure used on crops and from livestock waste (see AREI Part 4). Nine percent of domestic wells sampled by USGS's National Water Quality Assessment Program (NAWQA) during 1993-2000 had nitrate concentrations exceeding EPA's drinking water standard (maximum contaminant level or MCL) of 10 milligrams per liter, and agriculture was identified as the major source. EPA reported in its Water Quality Inventory that nutrient pollution is the leading cause of water quality impairment in lakes and a major cause of oxygen depletion in estuaries.

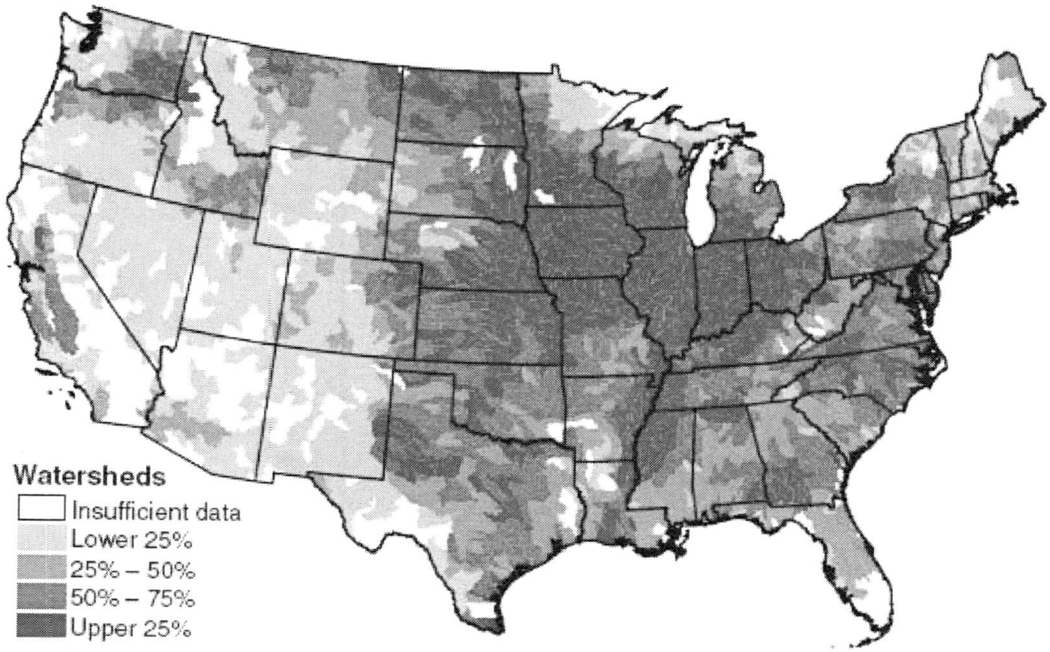

Source: Economic Research Service, USDA, based on erosion data from1997 National Resources Inventory, NRCS.

Figure 13. Potential delivery of sediment to surface water.

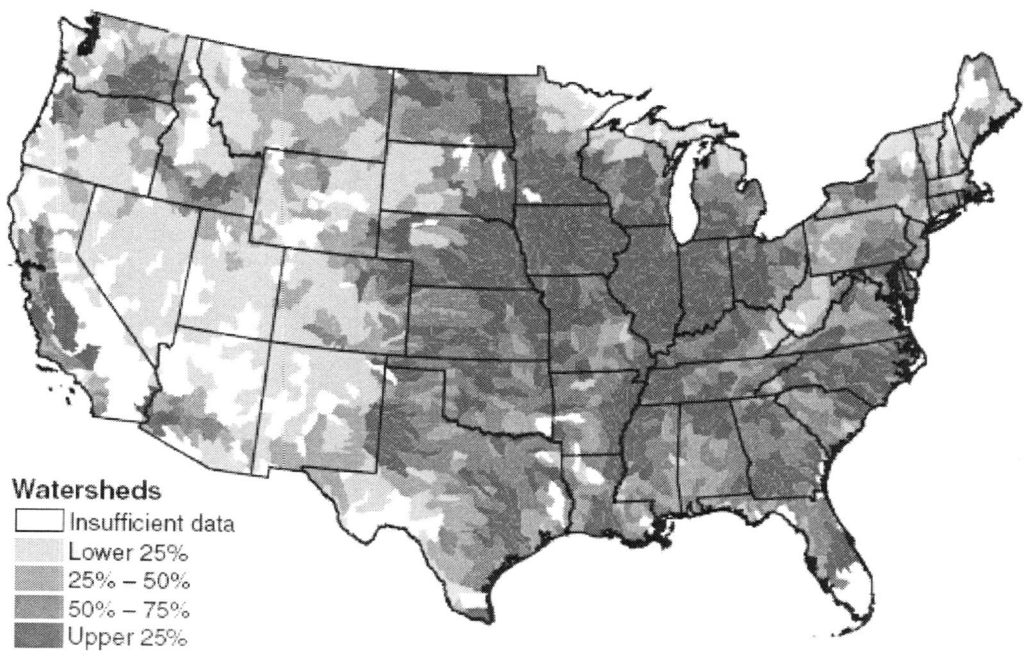

Source: Economic Research Service, USDA, based on nitrogen data from Association of American Plant Food Control Officials (1998) and Kellogg et al. (2000).

Figure 14. Potential delivery of nitrogen to surface water.

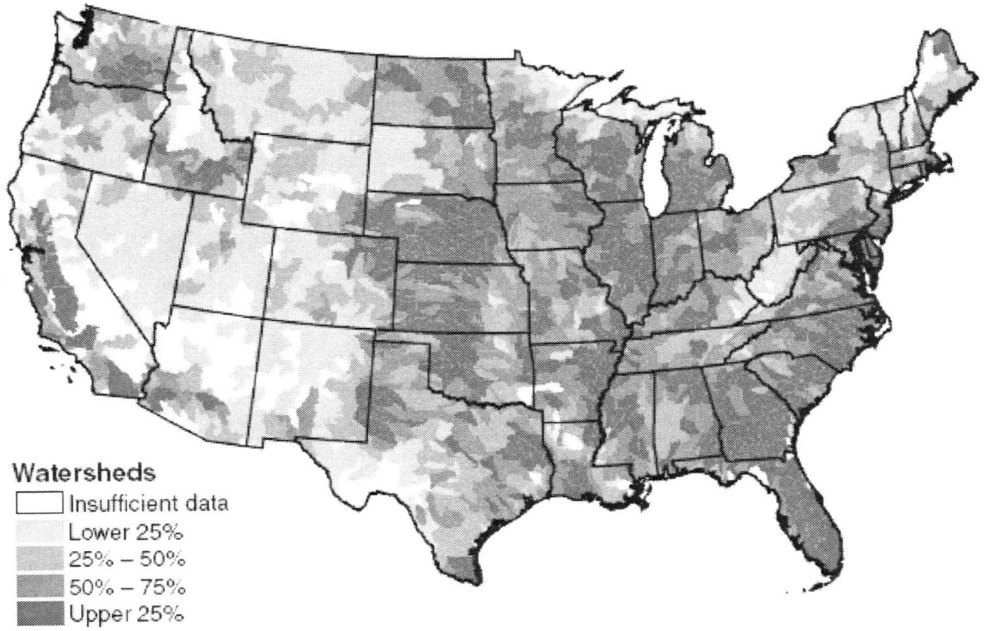

Source: Economic Research Service, USDA, based on nitrogen data from Association of American Plant Food Control Officials (1998) and Kellogg et al. (2000).

Figure 15. Potential nitrogen leaching to ground water.

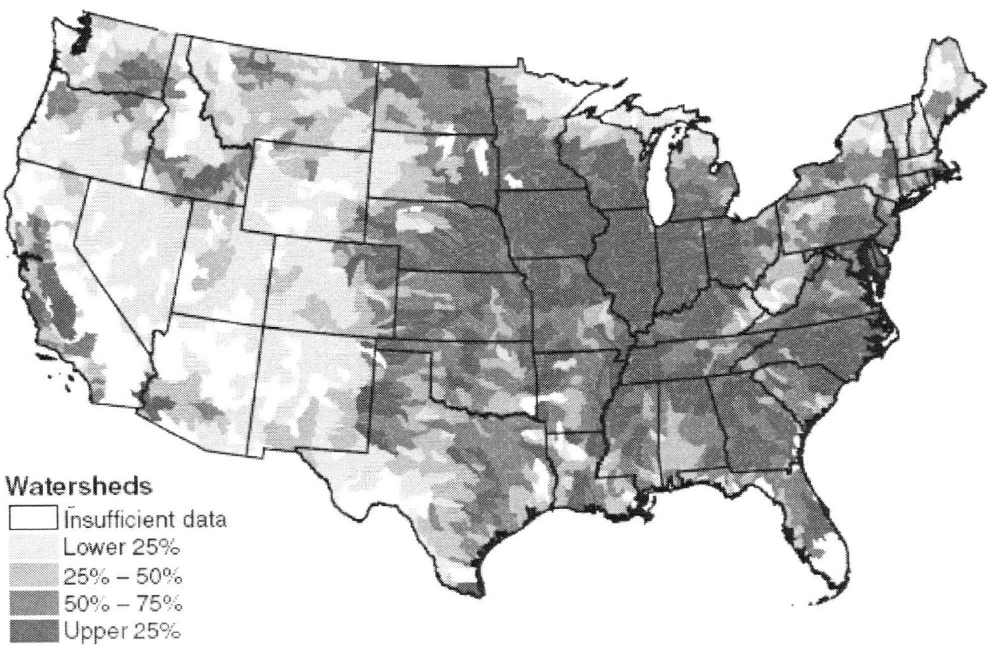

Source: Economic Research Service, USDA, based on phosphorus data from Association of American Plant Food Control Officials (1998) and Kellogg et al. (2000).

Figure 16. Potential delivery of phosphorus to surface water.

Watersheds with a high potential to deliver nitrogen to surface water are primarily in the Heartland and Southern Seaboard regions (Figure 14). Watersheds with a high potential to discharge nitrogen to ground water are primarily in the Southern Seaboard, Fruitful Rim, Heartland, and Prairie Gateway regions (Figure 15). Watersheds with a high potential to discharge phosphorus to surface water are located primarily in the Heartland, Southern Seaboard, and Northern Crescent regions (Figure 16).

Eutrophication and hypoxia (low oxygen levels) in the northern Gulf of Mexico have been linked to nitrogen loadings from the Mississippi River (NOS, NOAA, 1999). Agricultural sources (fertilizer, soil inorganic nitrogen, and manure) are estimated to contribute about 65 percent of the nitrogen loads entering the Gulf from the Mississippi Basin (Goolsby et al., 1999). As much as 15 percent of the nitrogen fertilizer applied to cropland in the Mississippi River Basin makes its way to the Gulf of Mexico.

The Gulf of Mexico is not the only coastal area affected by nutrients. Recent research by the National Oceanographic and Atmospheric Administration has found that 44 estuaries (40 percent of major U.S. estuaries) exhibit highly eutrophic conditions, caused primarily by nitrogen enrichment (Bricker et al., 1999). These conditions occur in estuaries along all coasts, but are most prevalent in estuaries along the Gulf of Mexico and Mid-Atlantic coasts. Watersheds with a high potential to discharge nitrogen from agriculture to estuaries are located primarily in the Heartland, Mississippi Portal, and Southern Seaboard regions.

Farmers apply a wide variety of **pesticides** to control insects (insecticides), weeds (herbicides), fungus (fungicides), and other problems (see Chapter 14, "Pest Management"). Well over 500 million pounds (active ingredient) of pesticides have been applied annually on farmland since the 1980s, and certain chemicals can travel far from where they are applied. Pesticide residues reaching surface-water systems may harm freshwater and marine organisms, damaging recreational and commercial fisheries. Pesticides in drinking water supplies may also pose risks to human health. At least one of seven prevalent herbicides was found in 37 percent of the groundwater sites examined by USGS as part of the National Water Quality Assessment Program, but all at low concentrations.

Watersheds with a high propensity to discharge pesticides to surface water are located primarily in the Heartland and Mississippi Portal regions (Figure 17). Watersheds with a high propensity to discharge pesticides to ground water are primarily in the Heartland, Prairie Gateway, and Southern Seaboard regions (Figure 18).

Some irrigation water applied to cropland may run off the field into ditches and to receiving waters (see AREI Parts 2 and 4). These irrigation return flows often carry **dissolved salts** as well as nutrients and pesticides into surface or ground water. Increased salinity levels in irrigation water can reduce crop yields or damage soils such that some crops can no longer be grown. Increased concentrations of naturally occurring toxic minerals—such as selenium, molybdenum, and boron—can harm aquatic wildlife and impair water-based recreation. Increased levels of dissolved solids in public drinking water supplies can increase water treatment costs, force the development of alternative water supplies, and reduce the lifespans of water-using household appliances.

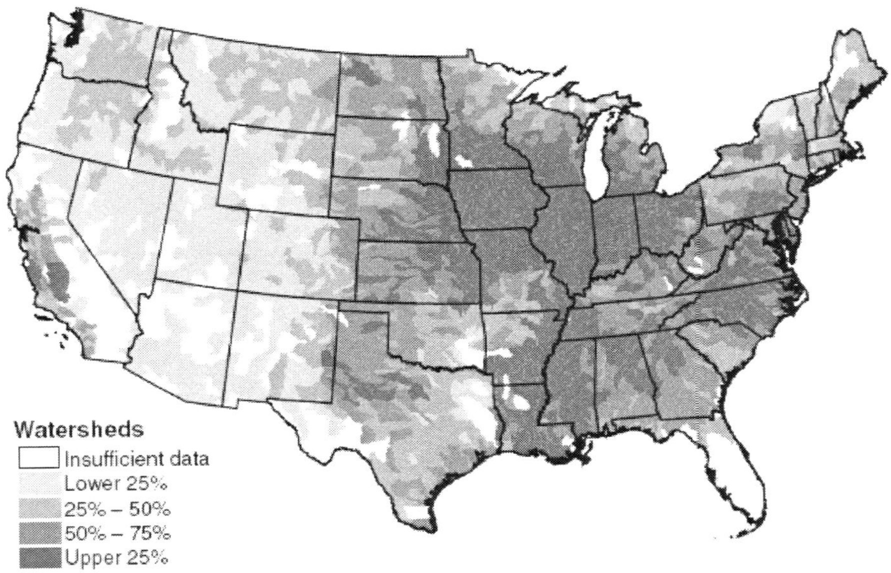

Source: Economic Research Service, USDA, based on pesticide data from USDA surveys.

Figure 17. Potential pesticide run off from cropland.

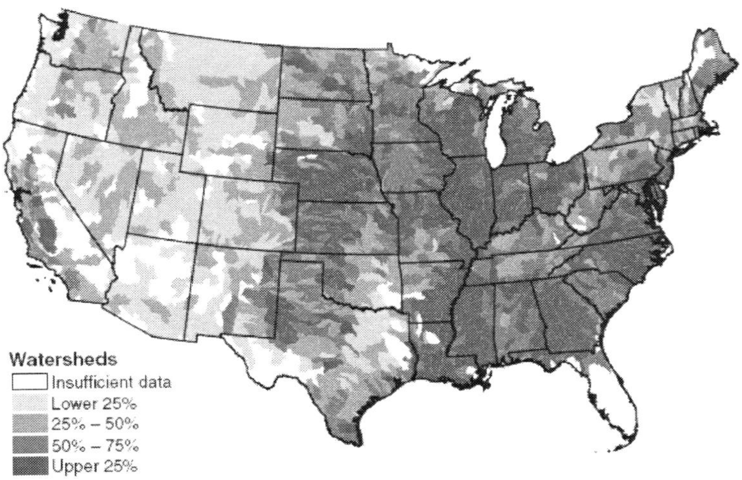

Source: Economic Research Service, USDA, based on pesticide data from USDA surveys.

Figure 18. Potential pesticide leading from cropland.

Dissolved salts and other minerals are a significant cause of pollution in the Prairie Gateway and arid portions of the Fruitful Rim and Basin and Range. Selenium is of particular concern because of its adverse biological effects. Selenium in irrigation return flows was identified as the cause of mortality, congenital deformities, and reproductive failures in aquatic birds in Kesterson Reservoir in western San Joaquin Valley, California (Seiler et al., 1999). A Department of Interior study of the Western United States found that 4,100 square miles of land irrigated for agriculture is susceptible to selenium contamination, along with adjacent land that may receive return flows (Seiler et al., 1999). Affected areas are primarily in California, western Kansas, eastern Colorado, and western South Dakota.

The possibility of **pathogens** contaminating water supplies and recreation waters is a continuing concern. Bacteria are the largest source of impairment in rivers and streams, according to EPA's water quality inventory. Potential sources include inadequately treated human waste, wildlife, and animal feeding operations (see Chapter 16, "Animal Agricultural and the Environment"). Diseases from micro-organisms in livestock waste can be contracted through direct contact with contaminated water, consumption of contaminated drinking water, or consumption of contaminated shellfish. Bacterial, rickettsial, viral, fungal, and parasitic diseases are potentially transmissible from livestock to humans (CAST, 1996). Fortunately, proper animal management practices and water treatment minimize this risk. However, protozoan parasites, especially *Cryptosporidium* and *Giardia*, are important sources of waterborne disease outbreaks. *Cryptosporidium* and *Giardia* may cause gastrointestinal illness, and *Cryptosporidium* may lead to death in persons with compromised immune systems. These parasites have been commonly found in beef herds and *Cryptosporidium* is widespread on dairy operations (USDA, APHIS, 1994; Juranek, 1995).

GOVERNMENT RESPONSE TO AGRICULTURAL POLLUTION

While agriculture's impacts on water resources are widespread and considered to be significant, the control of agricultural pollution is a challenge. The primary reason for this is that pollution from agriculture is generally "nonpoint" in nature. Nonpoint-source pollution has four characteristics that have an important bearing on the design of policies for reducing it.

- Nonpoint emissions are generated diffusely over a broad land area. These emissions leave from fields in so many places that it is generally not cost effective to accurately monitor emissions using current technology.
- Nonpoint emissions (and their transport to water or other resources) are subject to significant natural variability due to weather-related events and other environmental characteristics.
- Nonpoint emissions and the associated water quality impacts depend on many site-specific characteristics, such as soil type, topography, proximity to the water resource, climate, etc.
- Nonpoint pollution problems are often characterized by a very large number of nonpoint polluters.

The difficulties in measurement, variability of discharges, and the site-specific nature make regulations used for point sources (factories and sewage treatment plants) largely inappropriate for nonpoint sources. As a consequence, water quality laws such as the Clean Water Act (see Chapter 27, "Federal Laws Protecting Environmental Quality") generally do not regulate agricultural pollution but, instead, pass most of the responsibility on to the States. This has resulted in quite varied responses, reflecting the States'particular resource concerns and organizational capacity. Thirty-three States have laws with provisions that regulate agriculture under certain conditions, such as when voluntary approaches fail to achieve water quality goals. States commonly use technology standards that require farmers to implement conservation plans that contain recommended management practices (Ribaudo and Caswell, 1999), such as conservation tillage, nutrient management, pesticide management, and irrigation water management. These plans can be required statewide, or in areas particularly vulnerable to agricultural pollution.

By contrast, the Federal Government relies primarily on voluntary approaches, such as education and financial assistance (policy instruments), to encourage farmers to protect water quality. Major USDA programs such as the Environmental Quality Incentive Program and Conservation Security Program are important sources of information and assistance for farmers concerned with water quality (see Chapter 24, "Working-Land Conservation Programs").

Between 1997 and 2004, 37 percent of EQIP funds were devoted to water quality and conservation.

REFERENCES

[1] Association of American Plant Food Control Officials (1998). *Commercial Fertilizers, 1997.* University of Kentucky, Lexington, Kentucky.

[2] Bricker, S.B., C.G. Clement, D.E. Pirhall, S.P. Orlando, and D.R.G. Farrow (1999). *National Estuarine Eutrophication Assessment: Effects of Nutrient Enrichment in the Nation's Estuaries.* NOAA, Special Projects Office and the National Centers for Coastal Ocean Science. Sept.

[3] Council for Agricultural Science and Technology (CAST) (1996). *Integrated Animal Waste Management.* Task Force Report No. 128, November.

[4] Goolsby, D.A., W.A. Battaglin, G.B. Lawrence, R.S. Artz, B.T. Aulenbach, R.P Hooper, D.R. Keeney, and G.J. Stensland (1999). *Flux and Source of Nutrients in the Mississippi-Atchafalaya River Basin: Topic 3 Report.* Report submitted to White House Office of Science and Technology Policy, Committee on Environment and Natural Resources, Hypoxia Work Group. May.

[5] Juranek, D. (1995). "Cryptosporidiosis: Source of Infection and Guidelines for Prevention," *Clinical Infectious Diseases* 21(Suppl 1):57-61.

[6] Kellogg, R.L., C.H. Lander, D.C. Moffitt, and N. Gollehon (2000). "Manure Nutrients Relative to the Capacity of Cropland and Pastureland to Assimilate Nutrients: Spatial and Temporal Trends for the United States," USDA, Natural Resources Conservation Service, #NPS00-0579

[7] Koltun, G.F., M.N. Landers, K.M. Nolan, and R.S. Parker, (1997). "Sediment transport and geomorphology issues in the Water Resources Division." Proceedings of the U.S. Geological Survey Sediment Workshop, Feb. 4-7, Reston, VA, and Harpers Ferry, WV. U.S. Geological Survey, Reston, VA.

[8] National Ocean Service, National Oceanographic and Atmospheric Administration (1999). *Hypoxia in the Gulf of Mexico: Progress towards Completion of a National Assessment.*

[9] Ribaudo, M.O., and M. Caswell (1999). "U.S. Environmental Regulation in Agriculture and Adoption of Environmental Technologies," in *Flexible Incentives for the Adoption*

of Environmental Technologies in Agriculture, F. Casey, S. Swinton, A. Schmitz, and D. Zilberman (Editors), Kluwer Academic Press.

[10] Seiler, R.L., J.P. Skorupa, and L.A. Peltz (1999). *Areas Susceptible to Irrigation-Induced Selenium Contamination of Water and Biota in the Western United States.* USGS Circular 1180. U.S. Dept. Int., U.S. Geological Survey, Reston, VA.

[11] Smith, R.A., G.E. Schwarz, and R.B. Alexander (1994). *Regional Estimates of the Amount of U.S. Agricultural Land Located in Watersheds with Poor Water Quality.* Open-File Report 94-399. U.S. Dept. Int., U.S. Geological Survey, Reston, VA.

[12] Smith, R.A., R.B. Alexander, G.E. Schwarz, and M.C. Ieradi (2005). "Effects of Structural Changes in US Animal Agriculture on Fecal Coliform Contamination of Streams." Poster presentation, U.S. Dept. Int., U.S. Geological Survey, Reston, VA.

[13] U.S. Department of Agriculture, Animal and Plant Health Inspection Service (1994). *Cryptosporidium and Giardia in Beef Calves.* National Animal Health Monitoring System report. Jan.

[14] U.S. Environmental Protection Agency (2002). *National Water Quality Inventory: 2000 Report to Congress.* EPA841-R-02-001. Office of Water, Aug.

[15] U.S. Geological Survey (1999). *The Quality of Our Nation's Waters: Nutrients and Pesticides.* U.S. Geological Survey Circular 1225.

In: Agricultural Resources and Environmental Indicators ISBN 978-1-60021-467-7
Editor: Keith Wiebe and Noel Gollehon, pp. 49-57

Chapter 6

WETLANDS STATUS AND TRENDS

LeRoy Hansen

Almost half of all wetlands in the 48 contiguous States have been drained since colonial settlement—nearly 85 percent for agricultural use. Public policies that encourage wetland preservation began to emerge in the 1970s. The United States now appears to be achieving a goal of "no net loss" due, in part, to the conversion of some agricultural lands back to wetlands.

INTRODUCTION

Wetlands today cover over 7 percent of the nonfederal land area of the 48 States. Most wetlands occur on forest land, while 15 percent—about 16.6 million acres—occur on lands associated with crop production and pasture. Despite having a relatively small portion of total wetland acres, agriculture has played and is likely to continue to play a significant role in wetland policy (USDA, NRCS, 2004). For example, much of the past losses in wetlands are attributable to agriculture. Between 1954 and 2002, 66 percent of total wetland losses (24.2 million acres) were from converting to agricultural uses. Furthermore, future gains in wetlands will likely draw from agricultural lands. Between 1974 and 2002, over 50 percent of all lands converted to wetlands had been in agricultural use. With the continuation of private and public initiatives to restore wetlands, agricultural lands are likely to continue to be converted to wetlands because the conversion of agricultural lands is often less costly than conversion from other uses (like urban ones) (Heimlich et al., 1998).

Wetlands are a productive medium for forests, provide habitat for fish and wildlife, preserve water quality (see Chapter 5, "Water Quality: Impacts of Agriculture"), reduce flood damage, provide open spaces and recreational sites, and enhance wildlife diversity (Table 10).

For these reasons, society values wetlands. However, since most wetland benefits occur offsite, private owners usually cannot benefit economically from wetlands.

Society's awareness of the value of wetlands has grown only in the last several decades. Wetlands were once seen as "wasted" land that should be exploited. When colonists first set foot in America, there were 221-224 million acres of wetlands in what was to become the contiguous United States (there were another 170 million acres in Alaska and Hawaii, not discussed further here). Most of those wetlands were in three regions: the Midwestern States (27 percent), the Southeastern States (24 percent), and the Delta and Gulf States (24 percent). As settlement spread, wetlands were converted for other uses, with the pace increasing as available nonwetlands diminished and drainage technology improved (Heimlich et al., 1998).

Table 10. Wetland functions, services, and values

Wetland	Function	Service	Economic values
Private values			
Forest	Tree growth medium	Commercial timber harvest	Net economic value of timber
Fisheries	Fish habitat	Commercial fish harvest	Net economic value of commercial catch
Mixed values			
Recreation	Wildlife habitat	Recreational, fishing, and waterfowl harvest	Net economic value of hunting and fishing experience
Public values			
Flood control	Flood retention	Reduced flood flows/peak	Net economic value of reduced damages
Water quality	Water filtration	Cleaner waters	Net economic value of reduced damages
Endangered species	Endangered species	Biodiversity	Net option and existence values

Source: Adapted from Bergstrom and Brazee (1991).

WETLAND EXPLOITATION: SETTLEMENT TO 1954

Between the start of European settlement and 1954, 40-44 percent of original wetlands were drained or filled.[1] Most of this activity probably occurred after 1885, with as many as 80 million acres of wetlands and other areas drained by 1930 and, with a slowdown in conversions during the Depression and World War II, another 10-11 million acres drained between 1930 and 1954 (Figure 19).

With the explicit encouragement of Federal policies and local cooperative efforts, wetlands were converted to agricultural and other uses at an average net rate of 814,000 to 887,000 acres per year between 1885 and 1954.

Almost 30 percent of net wetland conversion during this period was in the Midwest, 22-24 percent in the Delta and Gulf region, and 14-16 percent in the Southeast (USDA, ERS, 2000). Data are insufficient to reveal gross changes from dryland to wetland, but some wetlands were probably restored or created as lands once converted were abandoned, drainage failed, and reservoirs or other impoundments saturated formerly dry land.

[1] Because we are not certain of the total wetland acreage prior to settlement, we provide range estimates of probable changes in wetland acres.

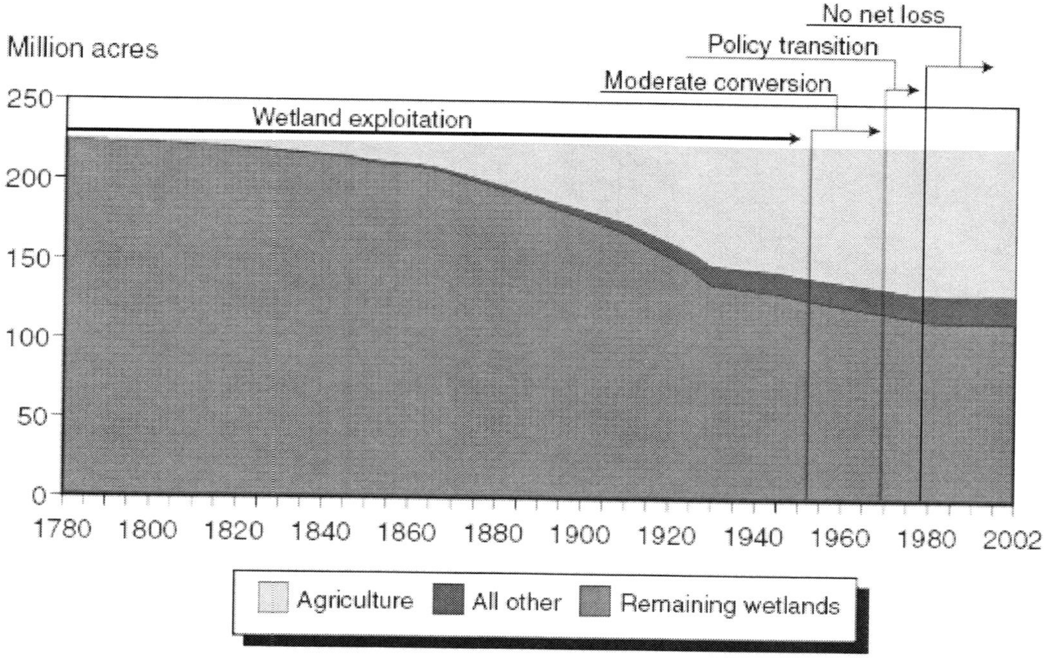

Source: ERS analysis of "Status and Trends in the Conterminous States: 1886-1997" (U.S. Dept. of the Interior's Fish and Wildlife Service), the 2002 National Resources Inventory (USDA, Natural Resources Conservation Service), and NRCS reported estimates at http://www.nrcs.usda.gov/news/archive/2004newsroom.html et al. (2000).

Figure 19. Status and losses of wetlands, 1780-2002.

MODERATE WETLAND CONVERSION: 1954-74

The pace of net wetlands conversion in 1954-74 was about half that of the long-term rate since settlement, dropping to an average of 458,000 acres per year (Figure 20).

Gross conversion to agriculture averaged 593,000 acres per year, while urban development, conversion to other uses, and water impoundments increased the total to 730,000 acres. Restoration of dryland and deep water to wetlands averaged 273,000 acres per year, about 1 acre restored for every 3 acres converted.

During this period, drainage shifted from the Midwest to the Delta and Gulf region (53 percent of all net conversion) and the Southeast (30 percent) (USDA, ERS, 2000). In the Delta, expansion for agricultural production in Louisiana, Mississippi, and Arkansas was probably the largest contributor to wetland conversion, although changes to coastal wetlands on the Louisiana gulf coast were also significant. In the Southeast, both urban and agricultural expansion in Florida and North Carolina were contributors. Net wetland acreage increased slightly in the Central Plains, Prairie Potholes, and Northeast, due to farmers'abandoning some agricultural land, increased rainfall expanding wetland area, and farmers'developing ponds and reservoirs on wetland fringes.

WETLAND POLICY TRANSITION: 1974-82

Growing public interest in wetland benefits during the 1970s resulted in Federal policy changes, such as the Clean Water Act's Section 404 and Executive Order 11990, which began to reduce wetland conversions. Section 404 (see Chapter 27, "Federal Laws Protecting Environmental Quality") established a program to regulate the discharge of dredged and fill material into waters of the United States, including wetlands. Landowners are required to obtain a permit from the Army Corps of Engineers to begin work in wetlands. The permit process involves a public review in which all interested parties can comment on potential adverse impacts from the proposed wetlands conversion (Alvayay and Baen, 1990; USEPA, 1993). Section 404(f) exempts some ongoing activities—including many farming, ranching, and forestry practices—when wetland impacts are expected to be minimal.

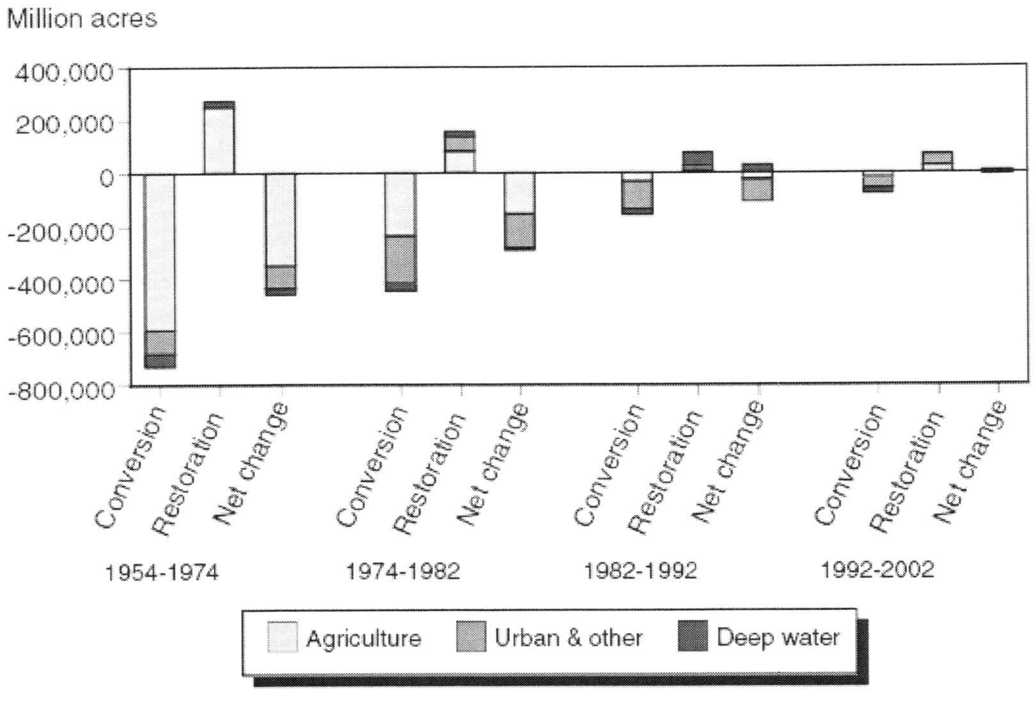

Source: ERS analysis of "Status and Trends in the Conterminous States: 1886-1997" (U.S. Dept. of the Interior's Fish and Wildlife Service), the 2002 National Resources Inventory (USDA, Natural Resources Conservation Service), and NRCS reported estimates at www.nrcs.usda.gov/news/archive/2004newsroom.html.

Figure 20. Changes in wetland acreage by use, contiguous States, 1954-2002.

Executive Order 11990, signed by President Carter in 1977, directed Federal agencies to minimize the loss and degradation of wetlands and, instead, to improve the health of wetlands. As a result, the 1974-82 rate of net wetland conversions dropped by 37 percent, to 290,000 acres per year, despite higher market prices for crops and greater economic incentives for agricultural conversion. Gross conversion for agriculture dropped to 235,000 acres per year, but a large increase in conversions to other uses left total gross conversion at

446,000 acres (Figure 20). Over this period, gross increases in wetlands fell to 156,000 acres per year, with agricultural lands accounting for more than half.

Wetland was converted primarily in the Southeast, which had more than 60 percent of net conversion in 1974-82, and the Delta and Gulf region, which had 30 percent (USDA, ERS, 2000). Three-fourths of Southeast conversions were North Carolina wetlands converted to agricultural land, while lost wetlands in coastal Louisiana and agricultural conversion in Mississippi and Texas contributed to net changes in the Delta region.

No Net Loss:1982-2002

The downward trend in the rate of wetland conversions continued in 1982-2002 due to several factors: swampbuster (see Chapter 23, Compliance Provisions for Soil and Wetland Conservation") provisions of the 1985 Food Security Act, more rigorous enforcement of Section 404 permitting, changes in income tax treatment of conversion investments, additional State regulations, and falling agricultural prices. The Wetland Reserve Program (WRP)—authorized in the 1990 Food, Agriculture, and Trade Act—has further reinforced wetland conservation (see Chapter 22, "Land Retirement Programs"). The WRP provides an easement payment and helps cover wetland restoration costs for cropland permanently converted back to a wetland.

Furthermore, a policy goal of "no net loss" of wetlands is also affecting changes in wetland losses, preservation, and restoration (White House, 1991; 1993). To date, the "no net loss" goal has been interpreted to mean wetlands should be conserved wherever possible, and that acres of wetlands converted to other uses must be offset through restoration and creation of wetlands, maintaining or increasing the wetland resource base (USDA, NRCS, 1995).

The antecedent of the "no net loss" goal in Federal wetlands policy was the National Wetland Policy Forum. The Forum's blue-ribbon panel of environmental, agricultural, business, academic, and government leaders concluded that "no net loss" was a reasonable goal:

> Although calling for a stable and eventually increasing inventory of wetlands, the goal does not imply that individual wetlands will in every instance be untouchable or that the "no net loss" standard should be applied on an individual permit basis—only that the nation's overall wetlands base reach equilibrium between losses and gains in the short run and increase in the long term. The public must share with the private sector the cost of restoring and creating wetlands to achieve this goal (Conservation Foundation, 1988).

Since initiation of no net loss, wetland area has begun to stabilize at around 134 million acres (Figure 19).

In 1982-92, net wetland losses fell to 79,000 acres per year—about 25 percent of the 1974-82 rate—and fell even further to 9,300 acres per year in 1992-2002 (Figure 20).

In 1982-92, wetland losses due to agriculture—at 31,000 acres per year—were only 20 percent of total gross conversions. Losses to agriculture fell to 19,000 acres per year in 1992-2002—about 25 percent of the total losses. In 1982-92, 57 percent of all wetland losses were due to urban development. The building boom of the 1980s may explain the urban conversion rate of 89,000 acres per year, which is seven times the 1974-82 rate.

Agriculture's role in wetland restoration appears to be growing. In 1982-92, agriculture supplied 10 percent (8,000 acres per year) of the restored wetland acreage. However, in 1992-2002, agriculture's contribution more than tripled to 28,000 acres per year—40 percent of the restored acreage.

The Nation may be reaching its goal of no net loss. The most recent data—1997 to 2002—show a net increase in wetlands of 13,800 acres per year, along with a 45-percent drop in wetland conversions relative to 1992-97.

BEYOND "NO NET LOSS"

On Earth Day 2004, the White House announced a new national goal—moving beyond "no net loss" of wetlands to an overall increase in wetlands and wetland quality. Specifically, the goal is to create, improve, and protect at least 3 million wetland acres over the next 5 years.

COSTS AND BENEFITS OF "NO NET LOSS"

The "no net loss" policy has preserved wetland functions when a wetland is left unchanged and increased functions when a wetland is restored. Both public programs (e.g., Water Bank, swampbuster, and Wetland Reserve Program)[2] and private organizations (e.g., the Nature Conservancy and the North American Wetland Conservation Fund) have successfully secured these benefits. The economic benefits of a wetland are difficult to measure because the number, type, and quality of functions vary, as does the number of people affected across wetland sites. As a result, per-acre values of wetland benefits range from a few dollars to $300,000 or more (USDA, ERS, 2001a). Note that the value of a wetland can be high when many people are affected, even though the value to each individual is relatively low.

The cost of "no net loss" is the opportunity cost of preserving or restoring a wetland. This cost equals the amount a firm or individual would pay for the right to convert a wetland—or the amount a landowner forgoes by not being able to sell the wetland for an alternative use. Swampbuster provisions (see Chapter 23, "Compliance Provisions for Soil and Wetland Conservation") limit farmers' ability to convert wetlands for agricultural uses, at an estimated average opportunity cost of $2,200 per acre—assuming that farmers are unable to sell wetlands for alternative uses and that swampbuster provisions will not expire (USDA, ERS, 2001b).

In recent years, the public sector (primarily through the WRP) and private organizations have purchased development rights to protect wetlands. Purchase costs range from several dollars per acre for wetlands with little potential for conversion to hundreds of thousands of dollars for wetlands near urban development (USDA, ERS, 2001b).

Acquiring rights to and restoring former wetlands can be less expensive than preserving a wetland. This is especially so when former wetland sites are marginally suited to economic uses—so that acquisition costs are low—and relatively easily restored. The acquisition costs

[2] In 1970, the Water Bank program became the first USDA program designed to temporarily protect wetlands.

associated with wetland restoration have averaged less than $800 per acre (USDA, ERS, 2001b).

Wetland Acres and Wetland Functions

Net increases in wetland acres do not ensure increases in wetland functions. Functions lost when a mature wetland is drained can be greater than those gained when a similar type of wetland is restored. The grassy depressional wetlands of the Northern Plains—the Prairie Potholes—can reach maturity within 5 years. Conversely, hardwood wetlands can take 30 years or more to mature. Some restored wetlands may never provide functions that match those provided before conversion. Reasons include the impact of historic and current land use activities in the surrounding landscape, lack of appropriate restoration techniques, landowner preferences for establishing a wetland subclass other than the one fitting the landscape, and site modifications to address adjacent landowner concerns with hydrologic restoration (USDA, NRCS, 2002).

Wetland Functions Tracked in the NRCS Pilot Study

— Static surface-water storage

— Dynamic surface-water storage

— Temporary surface-water storage

— Maintain characteristic static or dynamic storage, soil moisture, and groundwater interactions

— Provide environment for characteristic plant community

— Habitat structure within the wetland

— Habitat interspersion and connectivity among wetlands

— Nutrient cycling

— Removal of imported elements and compounds

— Retention of particulates

— Organic carbon export.

Source: Adapted from USDA, NRCS, 2002.

USDA's NRCS initiated a National Wetlands Functional Assessment Pilot in March 1998. The model used in the pilot addressed the relative capacity of wetlands to perform various ecosystem functions (see box "Wetland Functions Tracked in the NRCS Pilot Study").

The models were able to register a modest increase in mean levels of wetland functions of restored USDA program wetlands versus their former state as drained and cropped. However, the pilot was not intended as a comprehensive assessment of the functional condition of USDA conservation program wetlands (USDA, NRCS, 2002). Successes of the pilot study suggest that continued NRCS research is likely to produce better models of wetland functions that can aid future policy analyses and program design.

For additional information on wetlands, go to:

USDA/NRCS website: http://www.nrcs.usda.gov/technical/land/pubs/ib4text.html
EPA website: http://www.epa.gov/owow/wetlands/
Fish and Wildlife Service website: http://wetlands.fws.gov/
USGS website: http://www.nwrc.usgs.gov/

REFERENCES

[1] Alvayay, J., and J.S. Baen (1990). Wetland Regulation in the Real World. Washington, DC: Beveridge and Diamond, P.C.

[2] Bergstrom, J., and R. Brazee (1991). "Benefit Estimation," No Net Loss of Wetlands: What do Agricultural Economists have to Contribute? R.E. Heimlich (ed.), Staff Report No. AGES 9149, U.S. Dept. Agr., Econ. Res. Serv.

[3] The Conservation Foundation (1988). "Protecting America's Wetlands: An Action Agenda," Final Report of The National Wetlands Policy Forum. Washington, DC.

[4] Heimlich, R., K. Wiebe, R. Claassen, D. Gadsby, and R. House (1998). Wetlands and Agriculture: Private Interests and Public Benefits. AER-765. U.S. Dept. Agr., Econ. Res. Serv.

[5] Smith, R.D., A. Ammann, C. Bartoldus, and M.M. Brinson (1995). An approach for assessing wetland functions using hydro geomorphic classification, reference wetlands, and functional indices. Technical Report WRP-DE-9, U.S. Army Corps of Engineers, Waterways Experiment Station, Vicksburg, MS.

[6] U.S. Department of Agriculture, Economic Research Service (2001a). "Economic Values of Wetland Functions." ERS Briefing Room: Conservation and Environmental Policy: Questions and Answers.

[7] U.S. Department of Agriculture, Economic Research Service (2001b). "Costs of Conserving or Restoring Wetlands." ERS Briefing Room: Conservation and Environmental Policy: Questions and Answers.

[8] U.S. Department of Agriculture, Economic Research Service (2000). "Wetlands Remaining by Year and Wetland Region: 1780-1992." ERS Briefing Room: Conservation and Environmental Policy.

[9] U.S. Department of Agriculture, Natural Resources Conservation Service (2004). National Resources Inventory 2002 Annual NRI.

[10] U.S. Department of Agriculture, Natural Resources Conservation Service (2002). Assessing Wetland Functional Condition Change in Agricultural Landscapes. Wetland Technical Note No. 1.

[11] U.S. Department of Agriculture, Natural Resources Conservation Service (1995). Wetlands Values and Trends. RCA Issue Brief #4, Nov.

[12] U.S. Department of the Interior, Fish and Wildlife Service, 2000. "Status and Trends of Wetllands in Conterminous United States 1986 to 1997." By T.E. Dahl, Washington, DC. 82pp.

[13] U.S. Environmental Protection Agency (EPA) (1993). "Section 404 of the Clean Water Act: an Overview." Wetlands Fact Sheet #10. Office of Water, Office of Wetlands, Oceans, and Watersheds.

[14] White House Office of Environmental Policy (1993). Protecting America's Wetlands: A Fair, Flexible, and Effective Approach. Aug. 24.

[15] White House Office of the Press Secretary (1991). Fact Sheet: Protecting America's Wetlands. Press release dated Aug. 5.

PART 3: KNOWLEDGE RESOURCES AND PRODUCTIVITY

In: Agricultural Resources and Environmental Indicators ISBN 978-1-60021-467-7
Editor: Keith Wiebe and Noel Gollehon, pp. 61-70

Chapter 7

CROP GENETIC RESOURCES

Kelly Day-Rubenstein and Paul Heisey

Crop genetic resources are essential for agricultural production, and their use results in significant economic benefits. But conservation of crop genetic resources is complicated by their public goods attributes.

AGRICULTURE'S DEPENDENCE ON GENETIC RESOURCES

Agriculture and genetic resources are critically interdependent. All agricultural commodities, even modern varieties, descend from an array of wild and improved genetic resources from around the world. Furthermore, agricultural production depends on continuing infusions of genetic resources for yield stability and growth.

Genetic improvements have arisen in several ways. Before the development of modern varieties, farmers cultivated landraces. Landraces are varieties of crops that evolved and were improved by farmers over many generations. The pace of crop improvement accelerated as modern breeding techniques were developed that facilitated selection of specific desirable traits. Breeders have crossed different parental material and selected traits resulting in higher yields, quality changes, and desirable production traits.

Breeders have also sought resistance to pests and diseases, and tolerance to nonbiological stresses such as drought. Because pests and diseases evolve, breeders continually need new and diverse germplasm from outside the utilized stock, sometimes using wild relatives of cultivated crops and landraces, to find specific traits to maintain or improve yields (Duvick, 1986). USDA has estimated that new varieties are resistant for an average of 5 years, while it generally takes 8-11 years to breed new varieties (USDA, 1990). Plant breeders often rely on landraces or wild relatives as a last resort, because often it is more difficult to incorporate genetic material directly from these sources. Undesirable traits often accompany the trait of interest, and extensive breeding may be needed to produce a final variety. However, when used, genes from these materials have "often had a disproportionately large and beneficial impact on crop production" (Wilkes, 1991).

ECONOMIC VALUES OF GENETIC RESOURCES

Attaching a value to genetic resources is hard; describing their benefits is easier (Day-Rubenstein et al., 2005). The simplest value arises from the "direct use" of genetic resources to produce food and fiber or to help create new varieties of crops.

Conserved genetic resources may also have economic value even if they are not being used at the time. The option to exploit resources in the future, for uses not presently known, has considerable value, though this value is difficult to measure. Also, the information about a conserved resource has economic worth. For example, the fact that a species of potato occurring naturally in the Andes has genes adapted for high altitudes may guide breeders toward a set of related germplasm in the future.

Modern molecular biology techniques such as genomics hold promise for reducing the costs of searching for useful traits in conserved material, therefore increasing its value. At present, however, much work would be required to turn raw genetic sequence data into useful information (Attwood, 2000), and neither sequence data nor resources for sequencing are now available for landraces or wild relatives.

Various economic methods have been used to value genetic material, but isolating the contribution made by genetic resources is difficult. Breeders use the genetic material to create new varieties, but the research **effort** by breeders has value as well. Thus, many studies have focused on the value of "genetic enhancement," or the value arising from the *use* of genetic material by breeders.

For example, the Office of Technology Assessment (1987) estimated that genetic improvements have accounted for half the yield gains in major cereal crops since the 1930s. Thirtle (1985) estimated the contributions of biological advances to U.S. crop production, controlling for changes in other inputs such as fertilizers, machinery, and pesticides, and concluded that biological improvements contributed to 50 percent of the yield growth of corn, 85 percent for soybeans, 75 percent for wheat, and 24 percent for cotton. Duvick (2005) estimated that 50 percent of the increases in maize (corn) yields since the early 1930s have been due to breeding. To date, practically all published economic analyses of the collection of genetic material, conservation in gene banks, or use of genetic resources in plant breeding programs have shown significant economic benefits from these activities.

Besides estimating the total value of genetic improvements, it is also possible to estimate the distribution of these benefits. ERS researchers estimated the value of improved crop varieties by modeling the difference in economic welfare for both consumers and producers (crop and livestock) had there not been crop improvements in five major U.S. crops. U.S. producers generally gain as lower production costs outweigh the losses from lower commodity prices. Producer gains are estimated at over $160 million annually. Lower prices benefit consumers by an estimated $223 million per year. Together, the net economic effect from genetic enhancements is estimated at roughly $385 million per year. Economic welfare also rises worldwide. Consumer benefits from lower food prices outweigh producer losses, leading to net welfare gains estimated to exceed $600 million per year (Table 11).

GENETIC DIVERSITY

The loss of genetic diversity in a species, also called genetic erosion, has been identified in many commercially important crops. One reason for this decline in diversity has been the loss of landraces and wild relatives of cultivated crops. The loss of wild relatives occurs mainly through habitat conversion. Because the economic values of wild relatives can rarely be appropriated (i.e., captured) by landowners, they may have less incentive to preserve habitats for wild relatives than to devote land to alternative uses such as clearing for agricultural or urban use.

Table 11. Estimates of annual benefits from genetic enhancements in U.S. major crops

Region	Change in producer benefits	Changein consumer benefits	Total welfare change
		$ million	
United States	162	223	385
Canada	-17	18	1
European Union	-103	180	77
Other Western Europe	-10	16	6
Japan	-9	66	57
Australia/New Zealand	-14	8	-6
China/transitional economies	-171	210	39
Developing agricultural exporters	-61	62	1
Developing Asian importers	-5	14	9
Rest of world	-119	157	38
Total	-347	954	607

Source: Based on methodology used in Frisvold et al., 2003.

Genetic erosion of crop varieties can be hastened as landraces are displaced by commercially developed varieties. Farmers want high yield potential and desirable consumption attributes, and commercial varieties are often superior in these respects. While maintaining a diverse set of landraces may benefit plant breeding, individual farmers are unlikely to account for this when selecting seed. Landraces, though, become extinct if farmers stop planting and maintaining them.

Widespread adoption of genetically uniform crop varieties makes the crop population more susceptible to a widespread disease or pest infestation. Genetically uniform varieties may initially be more resistant to pests and diseases. But as pests and diseases evolve to overcome host plant resistance, genetic uniformity increases the likelihood that such a mutation will prove harmful to a crop; disease could affect newly vulnerable varieties accounting for a greater proportion of a crop's production. Genetic uniformity contributed to the spread of the Southern corn leaf blight, which reduced the U.S. corn crop by 15 percent in 1970. Since then, the genetic vulnerability of wheat and corn is thought to have lessened (in part because of efforts to breed in greater diversity), but the genetic uniformity of rice, beans, and many minor crops is still a concern (NRC, 1993; FAO, 1998).

Despite concerns that crop yields and production will become more variable (Swanson, 1996), yields for many major crops have been relatively stable. This is probably because temporal diversity (diversity through time) has replaced spatial diversity (diversity across an area) (Duvick, 1984). Modern plant breeding provides a steady release of new varieties with new traits for pest or disease resistance. Keeping ahead of pests and diseases through temporal diversity depends on the quality of germplasm in public gene banks and in private breeder collections. Many of the benefits of raw germplasm cannot be appropriated because genetic material has public good characteristics. As a result, private breeders rely on the public sector to collect, characterize, and perform pre-breeding enhancement of genetic materials to make them available for private use (Duvick, 1991).

TOOLS TO CONSERVE GENETIC RESOURCES—*IN SITU*

Most of the world's genetic diversity is found *in situ*. Species preserved in situ remain in their natural habitat. For agriculturally important species, the greatest diversity in landraces and in wild relatives may be found near their centers of origin, i.e., the places in which they were first domesticated (Figure 21). *In situ* preservation efforts, as well as germplasm collection activities for *ex situ* conservation, are often focused on centers of origin.

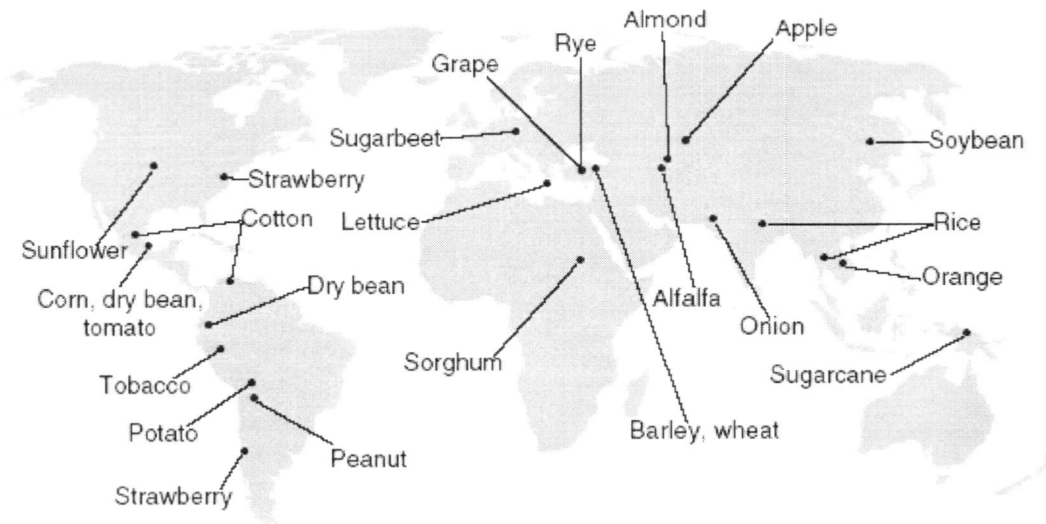

Source: GAO (1997).

Figure 21. Centers of origin, selected crops.

Because *in situ* conservation of agricultural genetic resources is carried out within the ecosystems of farmers'fields or wildlands, species continue to evolve with changing environmental conditions. *In situ* preservation can provide valuable knowledge about a species'development and evolutionary processes, as well as how species interact (Table 12).

In situ conservation of biodiversity is not more widely practiced because the private costs of doing so often outweigh the private benefits. Many decisions that affect conservation of biodiversity, such as choice of variety or deciding whether to clear land, are made at the individual or local level. To preserve agricultural genetic diversity *in situ*, a farmer may have to forgo a more profitable variety. For wild *in situ* resources, the land may need to be set aside completely.

It is difficult for countries—let alone individual farmers—to capture all of the value from genetic resources. Markets do not exist for most of the other environmental services provided by biological resources, such as benefits provided for wildlife species, and certain genetic resources are easy to transport and replicate.

Table 12. Advantages and disadvantages of in situand ex situconservation

In Situ conservation		Ex Situ conservation	
Advantages	**Disadvantages**	**Advantages**	**Disadvantages**
Genetic resources used to produce valuable product	Costs borne by farmers	Costs generally centralized	Certain types of germplasm not readily conserved
Evolutionary processes continue	May reduce farm productivity	Can preserve large amounts of diverse germplasm	Regeneration can be costly, time-onsuming
May better meet the needs of certain farms	Requires land	Germplasm can be more readily accessed by more breeders	Potential for genetic "drift" can reduce integrity of collection
More efficient for some germplasm, e.g., animals, crops that reproduce vegetatively	Farmer selections may not preserve targeted diversity	High-security storage impervious to most natural disasters	In practice, many collections are insufficiently funded, organized, and documented
Existing wild relatives can be preserved without collection			

Developing countries, where many *in situ* genetic resources for major crops are found, often face greater pressures for wildland conversion because of population growth and extensive farming techniques. In contrast, the quantity of agricultural land in the developed world has remained relatively stable or declined.

TOOLS TO CONSERVE GENETIC RESOURCES—*EX SITU*

The *ex situ* method of genetic resource conservation removes genetic material from its environment for long-term conservation, most often in gene banks. The world's gene banks presently hold more than 4 million accessions, or specific samples of crop varieties.

However, crop genetic resources must be collected, and only a fraction of the world's germplasm has been collected thus far. Stored plant materials must be kept under controlled conditions, and periodically regenerated (planted and grown) in order to maintain seed viability (Table 12). Not all kinds of plant genetic resources are easily conserved *ex situ*: some plants may need to be kept as living plants, a more costly process that requires additional land and labor. The resources necessary to maintain plant gene banks also face competing demands from other public programs.

U.S.Policies to Protect Genetic Resources

The United States promotes the conservation and use of genetic resources by (1) funding germplasm preservation efforts here and abroad and (2) pursuing international agreements. U.S. plant preservation is led by the National Plant Germplasm System (NPGS), which is administered by USDA's Agricultural Research Service. The NPGS, which houses more than 10,000 species, including wild relatives of crops, is one of the world's largest collectors and distributors of germplasm. It focuses on germplasm that may be needed by both public and private breeders, now and in the long term (see box, "Tyes of Germplasm"). Private incentives to collect and maintain such a collection are small, because any economic returns may not be realized until well into the future. Likewise, collecting exotic germplasm such as landraces and wild relatives can be expensive. However, it is a crucial source of needed traits, particularly resistance traits.

A recent study by the U.S. General Accounting Office found that relatively few wild relatives of domesticated varieties are held in gene banks, and not all collections have sufficient diversity (Table 13). Gene banks also may not be receiving adequate funding to fulfill their mission (Day, 1997). For example, the NPGS lacks sufficient funding to complete evaluation and documentation of its samples, or to perform necessary backups and regeneration of seed accessions (GAO, 1997).

International Policies on Genetic Resources

Most U.S. farmers produce non-native crops and livestock (NRC, 1993). Access to genetic resources in other countries is therefore critical to maintaining the rate of varietal improvement. Almost every plant species of major economic importance to the United States has been improved with germplasm from elsewhere. Past collection efforts and extensive breeding activities have resulted in the United States'actually being a net supplier of plant germplasm to the rest of the world (Figure 22). The NPGS supplies germplasm, free of charge, to anyone who requests it. Still, the United States continues to rely on other countries for genetic material. So, international agreements that affect the exchange of germplasm are an important tool for both U.S. policymakers and genetic resource managers.

The U.N. Convention on Biological Diversity (CBD), which came into force in 1993, is the most prominent international agreement addressing preservation of genetic resources. Historically, genetic material was regarded as the common heritage of humankind. Developing countries, the centers of origin for many crops, have often provided raw genetic material to public germplasm repositories.

Table 13. Some germplasm collections with insufficient diversity for reducing crop vulnerability

Collections with insufficient diversity to reduce crop vulnerability:

- Grapes
- Cool-season food legumes
- Sweet potatoes
- Cucurbits (e.g., cucumbers, squash, and pumpkins)
- Tropical fruit and nuts
- Walnuts
- Prunus (peach and cherry trees)
- Herbaceous ornamentals
- Woody ornamentals

Collections lacking specific types of germplasm:

- Wild and weedy relatives: almost 50%, including corn and soybeans.
- Landraces: 12 out of 40 collections, including corn, wheat, cotton, and alfalfa
- Genetic stocks: 50%, including alfalfa, peanuts, grapes
- Obsolete and current cultivars: 5 out of 40 collections

Source: GAO, 1997.

Types of Germplasm

Germplasm can be categorized into three basic types: (1) elite or modern, (2) landraces, and (3) wild and weedy relatives. Elite or modern germplasm has been improved by plant breeders. It may be a final cultivar (either recently developed or obsolete), or it may be germplasm that has been modified by a breeder for use in creating cultivars. Because landraces and wild or weedy relatives often contain unique traits, they increase the diversity of a germplasm collection. At the same time, elite material also contains diverse genes, which may be less exotic, but are generally easier to use (NRC, 1993). Thus, curators and breeders typically will want all three types of germplasm in a collection. In addition to these three basic types, germplasm collections also may include "genetic stocks," mutants and other germplasm with chromosomal abnormalities that are used by breeders.

Whether forgone earnings from raw genetic material are compensated for by free access to public genebanks and lower world food prices is an open question (Shands and Stoner, 1997; Fowler, 1991). But the traditional "free flow" of "unimproved" genetic resources and landraces between countries is no longer a given. The CBD is the most well-known in a serious of multilateral agreements to address (among other issues) ongoing disputes over the exchange and use of plant genetic resources. President Clinton signed the Convention in June

1993, but the U.S. Senate has not ratified it yet. The United States attends meetings as a non-voting observer.

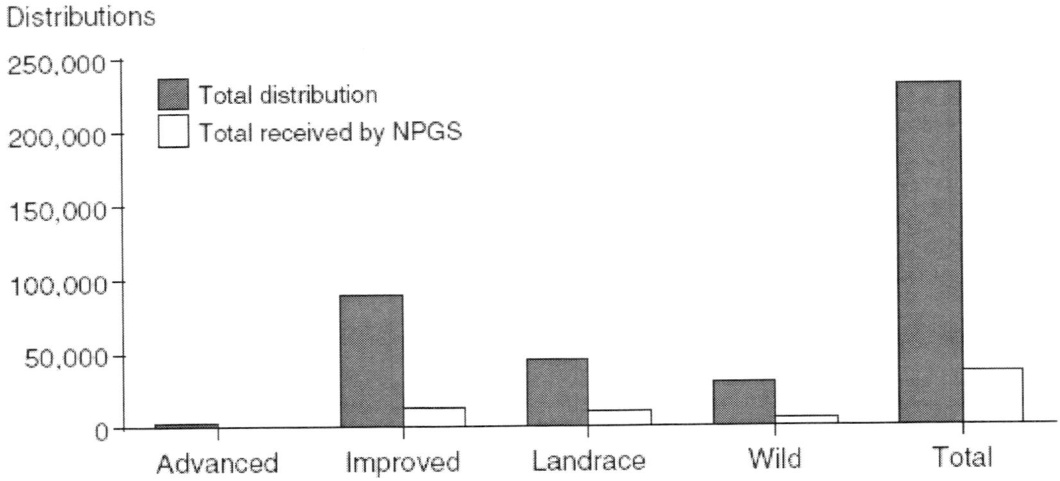

Source: National Germplasm Resources Laboratory, USDA.

Figure 22. National Plant Germplasm System: Distribution of germplasm, 1990-95.

In addition to the CBD, the International Treaty on Plant Genetic Resources for Food and Agriculture (IT) came into force in 2004. When terms are finalized, the treaty will govern the international exchange of germplasm for specified crops, including wheat, maize, rice, and alfalfa (though not other important crops such as soybeans, tomatoes, and peanuts). It is also intended as a mechanism to fund genetic resource conservation. As a result of the treaty, U.S. policymakers and genetic resource managers may soon face new exchange terms and rules governing benefit sharing. Many of the treaty's provisions are vague and uncertainties surround the valuation of crop genetic resources and the consequent sharing of benefits from germplasm preservation and exchange. The sources of funds for the preservation provisions of the treaty are also unclear.

The U.N. Convention on Biological Diversity (CBD), which came into force in 1993, is the most prominent international agreement addressing preservation of genetic resources. Historically, genetic material was regarded as the common heritage of humankind. Developing countries, the centers of origin for many crops, have often provided raw genetic material to public germplasm repositories.

Whether forgone earnings from raw genetic material are compensated for by free access to public genebanks and lower world food prices is an open question (Shands and Stoner, 1997; Fowler, 1991). But the traditional "free flow" of "unimproved" genetic resources and landraces between countries is no longer a given. The CBD is the most well-known in a serious of multilateral agreements to address (among other issues) ongoing disputes over the exchange and use of plant genetic resources. President Clinton signed the Convention in June 1993, but the U.S. Senate has not ratified it yet. The United States attends meetings as a non-voting observer.

In addition to the CBD, the International Treaty on Plant Genetic Resources for Food and Agriculture (IT) came into force in 2004. When terms are finalized, the treaty will govern the international exchange of germplasm for specified crops, including wheat, maize, rice, and alfalfa (though not other important crops such as soybeans, tomatoes, and peanuts). It is also intended as a mechanism to fund genetic resource conservation. As a result of the treaty, U.S. policymakers and genetic resource managers may soon face new exchange terms and rules governing benefit sharing. Many of the treaty's provisions are vague and uncertainties surround the valuation of crop genetic resources and the consequent sharing of benefits from germplasm preservation and exchange. The sources of funds for the preservation provisions of the treaty are also unclear.

The expansion of intellectual property rights may further affect genetic resource conservation and exchange. The CBD and IT establish property rights for plant germplasm in countries that are parties to the treaties, but the effects of these provisions on conservation have not yet been observed.

REFERENCES

[1] Attwood, T.K. (2000). "The Babel of Bioinformatics," *Science 290*, pp. 471-473.

[2] Day, P.R. (1997). "Biodiversity and the Equitable Use of the World's Genetic Resources" in *Global Genetic Resources: Access, Ownership and Intellectual Property Rights* (K.E. Hoagland and A.Y. Rossman, eds.) Beltsville symposia in Agricultural Research. Washington, DC: The Association of Systematics Collections.

[3] Day-Rubenstein, K., P. Heisey, R. Shoemaker, J. Sullivan, and G. Frisvold (2005). *Crop Genetic Resources: An Economic Appraisal.* U.S. Dept. Agr., Econ. Res. Serv., EIB-2, May.

[4] Duvick, D.N. (1984). "Genetic Diversity in Major Farm Crops on the Farm and in Reserve," *Economic Botany*, Vol. 38, pp. 161-178.

[5] Duvick, D.N. (1986). "Plant Breeding: Past Achievements and Expectations for the Future," *Economic Botany*, Vol. 40, No. 3, pp. 289-297.

[6] Duvick, D.N. (1991). "Industry and Its Role in Plant Diversity," *Forum for Applied Research and Public Policy*, Vol. 6, No. 3, pp. 90-94.

[7] Duvick, D.N. (2005). "The Contribution of Breeding to Yield Advances in Maize (zea mays l.)," *Advances in Agronomy*, Vol. 86, pp. 83-145.

[8] Fowler, C. (1991). "Stakes High in Battle for Genetic Diversity," *Forum for Applied Research and Public Policy*, Vol. 6, No. 3, 86-89.

[9] Frisvold, G., J. Sullivan, and A. Raneses (2003). "Genetic Improvements in Major US Crops: the Size and Distribution of Benefits," *Agricultural Economics*, Vol. 28, no. 2 (March), pp. 109-119.

[10] National Research Council (NRC) (1993). *Crop Diversity: Institutional Responses in Managing Global Genetic Resources: Agricultural Crop Issues and Policies.* Washington, DC: National Academy Press.

[11] Office of Technology Assessment, United States Congress (OTA) (1987). *Technologies to Maintain Biological Diversity*, OTA-F-330.

[12] Shands, H.L., and A.K. Stoner (1997). "Agricultural Germplasm and Global Contributions," in *Global Genetic Resources: Access, Ownership, and Intellectual Property Rights* (K.E. Hoagland and A.Y. Rossman, eds.). Washington, DC: The Association of Systematics Collections.

[13] Swanson, T. (1996). "Global Values of Biological Diversity: The Public Interest in the Conservation of Plant Genetic Resources of Agriculture," *Plant Genetic Resources Newsletter*, No. 105. Rome: International Plant Genetic Resources Institute.

[14] Thirtle, C.G. (1985). "Technological Change and the Productivity Slowdown in Field Crops: United States, 1939-78," *Southern Journal of Agricultural Economics*, 17 (Dec.) pp. 33-42.

[15] United Nations Food and Agricultural Organization (1998). *State of the World's Plant Genetic Resources for Food and Agriculture*. Rome.

[16] U.S. Department of Agriculture (USDA) (1990). "New Germplasm Helps Assure Food for the Future," USDA News Feature, Office of Public Affairs. Dec. 6.

[17] United States General Accounting Office (GAO) (1997). *U.S. Department of Agriculture: Information on the Condition of the National Plant Germplasm System.* Report to Congressional Committees.

[18] Wilkes, G. (1991). "*In Situ* Conservation of Agricultural Systems," *Biodiversity: Culture, Conservation, and Ecodevelopment* (J.B. Alcorn and M. Oldfield, eds.) Boulder, CO: Westview Press.

In: Agricultural Resources and Environmental Indicators
Editor: Keith Wiebe and Noel Gollehon, pp. 71-77
ISBN 978-1-60021-467-7

Chapter 8

AGRICULTURAL RESEARCH AND DEVELOPMENT

Margriet Caswell and Kelly Day-Rubenstein

Public and private research and development have driven impressive gains in agricultural productivity. Over the past few decades, advances in the biological sciences, as well as legislation that strengthened intellectual property protection for biological inventions, have increased research investment by the private sector. New institutional arrangements have fostered public and private collaboration in research, but it is unclear how industry consolidation and changes in public funding will affect agricultural research and its effects on productivity.

INTRODUCTION

Unprecedented growth in agricultural productivity over the past century can be attributed largely to investments in agricultural research and technology development (see Chapter 10, "Productivity and Output Growth in U.S. Agriculture"). Many developments—including more efficient agricultural machinery, agricultural chemicals and fertilizers, genetic improvements in crops, and changes in farm management techniques—have transformed U.S. agriculture. These developments have contributed to an abundant and affordable food supply for consumers.

Most early research efforts sought to replace increasingly expensive resources with less expensive ones. For example, the development of farm machinery helped offset increasing labor costs. Currently, demands for safer, healthier, and more convenient foods, natural resource conservation, environmental protection, and animal welfare are changing the agricultural research portfolio. These demands relate directly to agricultural products and to the impacts of production methods.

Research Demand

Many different forces affect research investment, and these forces differ for the public and private sectors. Some technology development is in response to consumer demand. This kind of focused research is often called "applied."The private sector will respond to market demands for new agricultural technologies, but markets may not address all external effects of production. Environmental regulation, for example, may increase the development of some environmentally benign technologies and the demand for those technologies.

Research can also be conducted without an immediately marketable product, usually for two reasons: basic research (to gain fundamental knowledge) and the provision of public goods. Basic research is conducted most often in the public sector because the results of the research lack immediate private payoffs. The results, though, can provide a scientific foundation for later public and private developments. Developments in biotechnology have blurred the distinction between basic and applied research. For example, "theoretical" fields such as genomics, proteomics, and bioinformatics have been supported strongly by the private sector.

Public goods represent a market failure because an individual's use of the good does not diminish its availability to others, and it is difficult to exclude anyone from using the good. National defense exemplifies a public good because once security is provided for one, all receive the same protection. In agriculture, food safety and ecosystem stewardship have public good characteristics. While the payoff to society of investing in basic and public good research is high, the results of such research generally cannot be appropriated, so the private sector has little market incentive to conduct this research. That is where government steps in—through funding and technology transfer activities.

The roles of the public sector and private industry in agricultural research have undergone significant changes in the last two decades due to developments in science, policy, and markets. The public sector was the primary investor in agricultural research prior to the 1980s, but now the private sector funds the development of many new agricultural technologies (Fuglie et al., 1996; Huffman and Evenson, 1993; Klotz et al., 1995; and Pray, 1993) (see Figure 23).

Public Sector Research and Development

Public agricultural research involves a unique partnership between the Federal Government (chiefly USDA) and the States. USDA, the State Agricultural Experiment Stations (SAES), and cooperating institutions together conducted over $4 billion of research in 2002 (USDA Current Research Information System). USDA conducts much of its inhouse research through its research agencies, primarily the Agricultural Research Service, the Forest Service, and the Economic Research Service. The largest expenditures on agricultural research in the public sector are made by SAES and cooperating institutions, which rely on Federal and State funding, as well as the private sector.

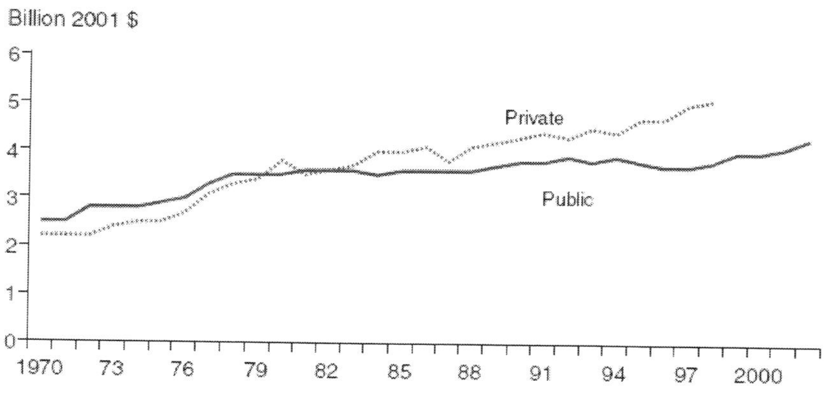

Source: USDA, Economic Research Service.

Figure 23. Agricultural R&D expenditures, 1970-2002.

Historically, USDA has used several funding instruments to provide research money to States. *Formula funds* are allocated in block form to States based on rural population and number of farms. Research administrators have numerous options in how they distribute formula funds. *National Research Initiative (NRI) competitive grants* are allotted by peer review panels. Special grants are awarded by Congress, whereas other USDA *contracts, grants, and cooperative agreements* are awarded at the discretion of USDA research agencies. (See Fuglie et al., 1996; and National Research Council, 1996, for descriptions and comparisons of these mechanisms.)

Within the public agricultural research sector, natural resource and environmental issues are of interest because they have both local and national dimensions. State research investments might be focused on local problems, with Federal funds earmarked for larger geographic issues. For example, the development of technologies to improve water quality and increase water-use efficiency can have critical local benefits (see AREI Part 2). However, benefits from improved water quality accrue beyond regional jurisdictions. Overall, public research on natural resources and the environment accounted for 21 percent of total public agricultural funds in 2003, up from 17 percent in 1998 (Figure 24).

The research categories that we use may not capture all research that can benefit the environment. Scientists self-classify their research using USDA's Current Research Information System (CRIS) and may not consider "natural resource and environmental research" as the primary objective of their work. For example, plant breeders may produce resistant varieties that require fewer agricultural chemicals, which may improve water quality. Still, they may classify the research under "plants and their systems."

PRIVATE SECTOR RESEARCH AND DEVELOPMENT

Private industry has been playing a more important role in agricultural research, not only boosting research investments but also expanding into new areas of research. For more information, see the "Agricultural Research and Productivity" Briefing Room. Private industry expenditures on agricultural research have increased 50 percent in real terms between 1978 and 1998[1] (Figure 23). In 1998, 60 percent of private sector agricultural research expenditures were allocated to biological and chemical technologies, such as agricultural chemicals, plant breeding, and animal health, compared with only 19 percent in 1960 (Figure 25).

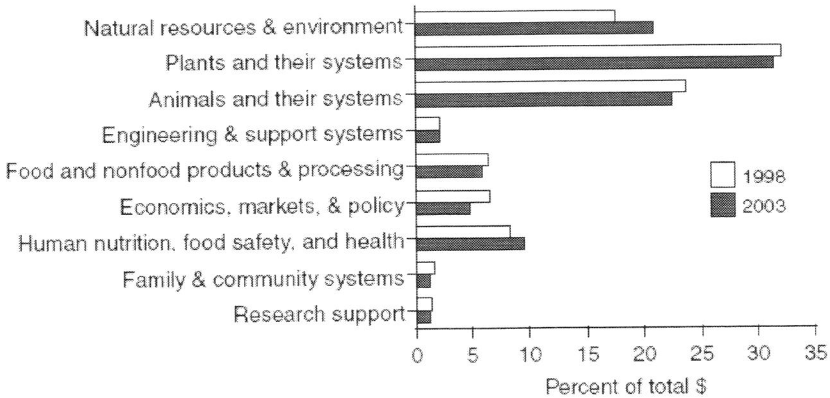

Source: USDA Current Research Information System.

Figure 24. Allocation of public funds for agricultural research, 1998 and 2003.

Advances in the biological sciences and expanded intellectual property rights (IPRs) protection for biological innovations have stimulated private sector efforts in technology development. Basic research in biology, microbiology, and computing created new technological opportunities for private agricultural research. For example, gene transfer technologies enable researchers to tailor crops for specific uses, such as crops resistant to disease, pests, herbicides, or harsh environmental conditions; and crops with increased nutrition or improved food processing traits. (See Chapter 9, "Biotechnology and Agriculture" for a more complete discussion of biotechnology-derived agricultural innovations)

Expanded IPRs for biological inventions and new plant varieties have allowed innovating firms to capture a greater share of the benefits from research. The Patent Act of 1790 was established to "promote the progress of science and useful arts," but biological inventions were considered products of nature at that time, and were not thought to be patentable. The extension of IPRs to new plant varieties and biological inventions, including biotechnologies, has further stimulated private companies to invest in plant breeding. The Plant Patent Act of 1930 and the Plant Variety Protection Act (PVPA) of 1970 established plant breeders'rights for new plants and plant varieties. In 1980, a Supreme Court decision (Diamond v. Chakrabarty) established the use of Utility Patents for biological inventions, specifically microorganisms. Further decisions by the Patent and Trademark Office broadened the use of Utility Patents for plants (in *ex parte* Hibberd in 1985) and animals (in *ex parte* Allen in 1987). The number of plant patents, Plant Variety Protection Certificates (PVPCs), and utility patents issued over the last 30 years has risen (Figure 26). International organizations have attempted to harmonize intellectual property protection in order to facilitate trade and technology development.

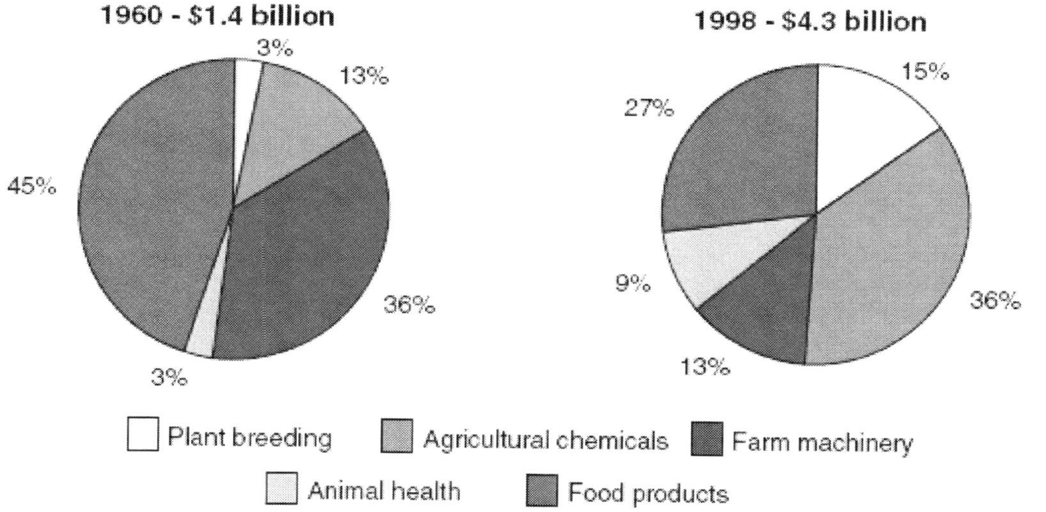

Figure in 1996 dollars.
Source: USDA, Economic Research Service.

Figure 25. Private agricultural research by industry, 1960 and 1998.

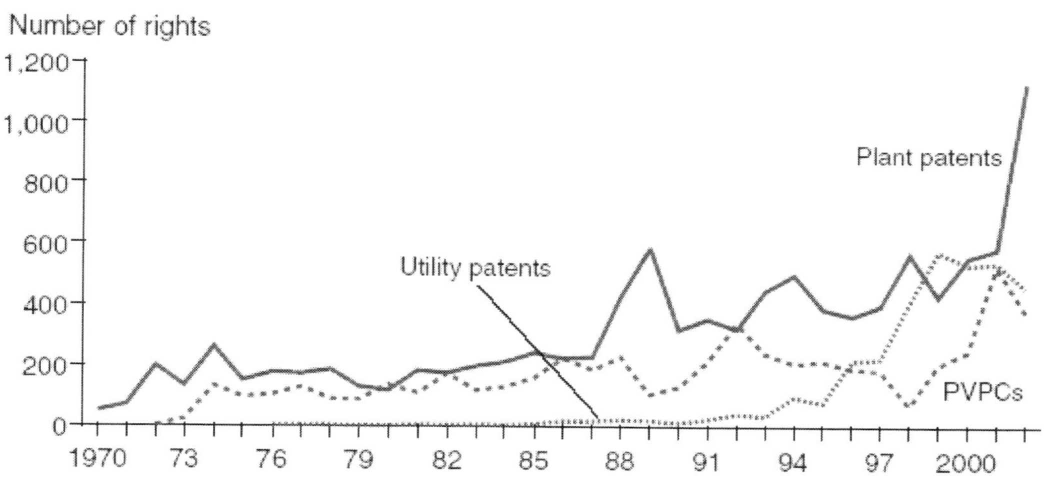

Source: USDA, Economic Research Service.

Figure 26. Intellectual property rights issued for new plant varieties.

PUBLIC AND PRIVATE COLLABORATION IN AGRICULTURAL RESEARCH AND TECHNOLOGY TRANSFER

Another change affecting technology development in agriculture has been the growth in collaborations between the public and private sectors. Before 1980, U.S. patent policy limited collaboration between public and private researchers, since the Federal Government assumed ownership of any inventions that resulted from federally funded research. The Government Patent Policy of 1980 (Bayh-Dole Act) granted institutions "certainty of title" for inventions resulting from federally funded research, and allowed Federal laboratories to issue exclusive licenses for patents of their inventions. The 1980 Stevenson-Wydler Technology Innovation Act mandated that each Federal research agency develop specific mechanisms for disseminating government innovations. The 1986 Technology Transfer Act gave government agencies additional means to foster technology transfer by authorizing public-private Cooperative Research and Development Agreements (CRADAs). This mechanism allows USDA to share technologies at various stages of development, research results, and scientific resources (though not money) with industry through joint research ventures.

Incentives for technology transfer may be very important, particularly for innovations that provide public-good benefits. Potential technologies developed in the public sector are not automatically marketed by the private sector. USDA and the SAES transfer a variety of innovations to private firms and directly to farmers, both shielded and unshielded (i.e., protected by IPRs or not) to ensure the provision of useful technologies to the agricultural sector (Day-Rubenstein and Fuglie, 2000).

Public entities like USDA can patent inventions meeting the criteria of the U.S. Patent and Trademark Office, then grant an exclusive/co-exclusive (most often), limited exclusive, or nonexclusive license to a private company to use or market the invention. In 2000, licensing revenue was less than 0.5 percent of USDA's R&D budget. Still, the licenses offer an incentive to private firms to develop and deploy the new technologies.

Other forms of cooperative effort between research entities include research consortia, which bring together several institutions to undertake joint research. These consortia increase funding support for strategic research and research that is considered to be long term and high risk (Fuglie and Schim-melpfennig, 2000). Large-scale efforts in plant genomics are underway to map, sequence, and analyze the genomes of several model plant species that are important for developing new crop varieties with desired traits.

Likely Research Trends

Several developments will influence the research portfolio over the next decade. Markets are beginning to develop for some public goods, such as products grown with "environmentally friendly" agricultural practices. If private firms can profit from providing products with desired social characteristics, research will accommodate such trends.

Another development that may affect future R&D investments is recent consolidation of seed, biotechnology, and agricultural chemical industries (Fernandez-Cornejo, 2004). There were 381 mergers, acquisitions, and other strategic alliances in the agricultural input industry between 1980 and 1998, and 10 firms accounted for almost half of that activity (King, 2001). Increased market power resulting from industry concentration and increased appropri-ability of technology may enhance incentives for private-sector innovation, leading to greater agricultural productivity. On the other hand, too much market power may inhibit technological advancement by creating barriers to entry for new firms and limiting access to critical technology and knowledge.

Developments in multiple scientific disciplines have led to several new fields: bioremediation, nanotechnology, genomics, proteomics, and bioinfor-matics. The expanded platform of knowledge will increase the options for agricultural research, development, and technology transfer.

REFERENCES

[1] Agricultural Marketing Service (various years). Plant Variety Protection Office.

[2] Agricultural Research Institute (1985). *A Survey of Agricultural Research by Private Industry III*. Bethesda, MD.

[3] Day-Rubenstein, Kelly, and Keith Fuglie (2000). "The CRADA Model for Public-Private Research and Technology Transfer in Agriculture," *Public-Private Collaboration in Agricultural Research: New Institutional Arrangements and Economic Implications*. Keith Fuglie and David Schimmelpfennig (eds.). Iowa State University Press, Ames, IA.

[4] Fernandez-Cornejo, J. (2004) *The Seed Industry in U.S. Agriculture: An Exploration of Data and Information on Crop Seed Markets, Regulation, Industry Structure, and Research and Development*. AIB-786. U.S. Dept. Agr., Econ. Res. Serv. Feb.

[5] Fuglie, Keith, Nicole Ballenger, Kelly Day, Cassandra Klotz, Michael Ollinger, John Reilly, Utpal Vasavada, and Jet Yee (1996). *Agricultural Research and Development: Public and Private Investments Under Alternative Markets and Institutions*. AER-735. U.S. Dept. Agr., Econ. Res. Serv. July.

[6] Fuglie, Keith, and David Schimmelpfennig, editors. (2000). *Public-Private Collaboration in Agricultural Research: New Institutional Arrangements and Economic Implications*. Iowa State University Press, Ames, IA.

[7] Huffman, Wallace, and Robert Evenson (1993). *Science for Agriculture: A Longterm Perspective*. Iowa State University Press, Ames, IA.

[8] King, John (2001). *Concentration and Technology in Agricultural Input Industries*. AIB-763. U.S. Dept. of Agr., Econ. Res. Serv. March.

[9] Klotz, Cassandra, Keith Fuglie, and Carl Pray (1995). "Private-Sector Agricultural Research Expenditures in the United States, 1960-92." Staff Paper No. AGES9525. U.S. Dept. Agr., Econ. Res. Serv. Oct.

[10] National Research Council (1996). *Colleges of Agriculture at the Land Grant Universities: Public Service and Public Policy*. National Academy Press, Washington, DC.

[11] Patent and Trademark Office (2001). Data from online patent search.

[12] Pray, Carl (1993). "Trends in Food and Agricultural R&D: Signs of Declining Competitiveness?" *U.S. Agricultural Research: Strategic Challenges and Options*. Robert D. Weaver (ed.), Agricultural Research Institute, Bethesda, MD, pp. 51-67.

[13] Pray, Carl, and Keith Fuglie (2000) "The Private Sector and International Agricultural Technology Transfer," *Public-Private Collaboration in Agricultural Research: New Institutional Arrangements and Economic Implications*. Keith Fuglie and David Schimmelpfennig (eds.). Iowa State University Press, Ames, Iowa.

[14] U.S. Department of Agriculture (various years, 1979-2001). Current Research Information System.

In: Agricultural Resources and Environmental Indicators ISBN 978-1-60021-467-7
Editor: Keith Wiebe and Noel Gollehon, pp. 79-89

Chapter 9

BIOTECHNOLOGY AND AGRICULTURE

Jorge Fernandez-Cornejo

Farmers adopting first-generation genetically engineered (GE) crops derive tangible benefits, even though not all benefits are reflected in standard measures of net returns. The impacts of GE crops vary with annual pest infestations, seed premiums, prices of alternative pest control programs, and any premiums paid for segregated (i.e., non-GE) crops.

INTRODUCTION

The unprecedented growth in crop yields and agricultural productivity over the 20th century owes much to a series of biological innovations embodied in seeds, beginning with the development of hybrid crops in the United States in the early part of the century and continuing with high-yielding varieties during the Green Revolution of the 1960s and 1970s. More recently, developments in modern biotechnology are expanding the processes of biological innovations by providing new tools. Agricultural biotechnology is a collection of scientific techniques, including genetic engineering, that are used to create, improve, or modify plants, animals, and microorganisms. Genetic engineering (GE) techniques allow a more precise and time-saving alteration of a plant's traits (facilitating the development of characteristics not possible through traditional plant breeding), and permit targeting of a single plant trait (decreasing the number of unintended characteristics that may occur with traditional breeding). Despite the benefits, however, environmental and consumer concerns currently limit acceptance of agricultural biotechnology, particularly in Europe. The ultimate contributions of agricultural biotechnology will depend on our ability to recognize its potential benefits and its risks (Fernandez-Cornejo at al., 1999).

Despite a focus here on genetically engineered crops in agriculture, the future importance of genetically engineered animals should not be understated. As a National Research Council (NRC) report indicates, the increased demand for meat and deterioration and loss of agricultural land will lead to pressures to exploit biotechnology to improve productivity in animal agriculture.

Table 14. Biotech crops currently available and in development ("in the pipeline") in the U.S.

Crop	Input traits					
	Herbicide tolerance	Insect resistance	Virus/fungus resistance	Agronomic properties[9]	Product quality[11]	Other[13]
Corn	C	C[5]	D	D	D	D
Soybeans	C	D	--	D	D	--
Cotton	C	C[6]	--	D	D	--
Potatoes		C[7]	D	D	D	D
Wheat	C[2]	--	D	--	--	--
Other field crops[1]	C[3] D[4]	D	D	D	D	D
Tomato, squash, melon	--	--	D	D	C[12] D	D
Other vegetables	D	--	--	--	D	--
Papaya	--	--	C[8]	--	--	--
Fruit trees	--	--	D	--	D	--
Other trees, flowers	--	--	--	D[10]	D	--

C = Currently available; D = In various stages of development.

[1] Includes barley, canola, peanuts, tobacco, rice, alfalfa, etc.

[2] Monsanto discontinued breeding and field-level research on its Roundup Ready wheat in 2004, deferring all further efforts to introduce it.

[3] Canola.

[4] Barley, rice, sugarbeets.

[5] Bt corn to control the corn borer commercially available since 1996; Bt corn for corn rootworm control commercially available since 2003.

[6] Bt cotton to control the tobacco budworm, the bollworm, and the pink bollworm commercially available since 1996.

[7] Bt potatoes resistant to the Colorado potato beetle commercially introduced in 1996. They were withdrawn from the market in 1999.

[8] Researchers at Cornell University and at the University of Hawaii developed two virus-resistant varieties of GE papaya. First commercial plantings were made in 1998. They were successful and were planted on more than 30 percent of Hawaii's papaya acreage in 1999.

[9] Resistance to cold, drought, frost, salinity; more efficient use of nitrogen; increased yield.

[10] Modified lignin content.

[11] Includes delayed ripening (fruits and vegetables with longer shelf life); increased protein, carbohydrate, and oil content; improved fiber properties (cotton), gluten content (wheat), naturally decaffeinated (coffee).

[12] Tomato genetically engineered to remain on the vine longer and ripen to full flavor after harvest was withdrawn from the market.

[13] Includes nutraceuticals, pharmaceuticals, and industrial products, such as increased vitamin, iron, beta-carotene (antioxidant), lycopene (anti-cancer), amino acid content; antibodies; vaccines; and specialty machine oils.

Sources: Virginia Polytechnic Institute and State University; USDA, APHIS; Colorado State; Shoemaker et al.; Pew.

GE crops are often classified into one of three generations (Panos, 1998). First-generation crops have enhanced input traits, such as herbicide tolerance, insect resistance, and resistance to environmental stresses like drought. Second-generation crops have added-value output traits, such as nutrient-enhanced seeds for feed. Third-generation crops produce pharmaceuticals, bio-based fuels, and products beyond traditional food and fiber (Table 14). At present, GE crops widely adopted are first-generation.

Seed Industry

Until the 1930s, most commercial seed suppliers were small, family-owned businesses lacking the financial resources to pursue their own research and development. These small businesses depended almost exclusively on plant breeding research in the public sector. The development and rapid producer acceptance of hybrid corn and greater legal protection of intellectual property rights brought large-scale change to the seed industry, particularly rapid increases in private R&D and market concentration in the U.S. seed industry.

Private R&D expenditures on plant breeding increased 1,300 percent between 1960 and 1996 (adjusted for inflation), while real public R&D expenditures changed little (Fernandez-Cornejo, 2004a, Figure 14). Two principal forms of legal protection behind the growth in private R&D on crop varieties are plant variety protection (PVP) certificates issued by the Plant Variety Protection Office of the USDA and patents issued by the U.S. Patent and Trademark Office of the U.S. Department of Commerce. Ag biotech patents, mostly dealing with some aspect of plant breeding, have outpaced the general upward trend in patenting throughout the U.S. economy. During 1996-2000, 75 percent of over 4,200 new agricultural biotechnology patents went to private industry. As private R&D on plant breeding grew rapidly, market concentration also increased. For example, the four largest corn seed firms accounted for nearly 70 percent of U.S. corn seed sales in 1997, and the four largest cotton seed firms provided more than 90 percent of the cotton seed varieties planted (Fernandez-Cornejo, 2004a, pp. 30-37). For more on R&D, see Chapter 8, "Agricultural Research and Development."

Biotech R&D

The creation of new plant varieties with useful agronomic properties requires significant knowledge of traditional plant breeding. Moreover, the commercial success of GE crop varieties typically requires that biotechnology-derived trait enhancements be incorporated into successful cultivars. In this sense, plant breeding and biotechnology are complementary. Acquisition of firms with established varieties by companies with the ability to improve varieties using biotechnology is one possible rationale for recent consolidation in the U.S. seed industry.

The number of field releases of plant varieties for testing purposes provides a useful indicator of R&D efforts on GE crops. The release of GE varieties of organisms into the environment is regulated and monitored by USDA's Animal and Plant Health Inspection Service (APHIS). Private companies and public institutions proposing tests of such organisms in the environment either notify APHIS of their intent or submit an application for a field release permit (referred to here as an application). If an APHIS review of the application (notification or permit application) establishes that there are no significant environmental risks associated with a release, a notification is acknowledged or a field permit is issued (referred to here as an "approval").

The number of applications received by APHIS for GE plant varieties increased from 9 in 1987 to a high of 1,206 in 1998. By mid-February 2005, nearly 11,300 applications had been received and more than 10,400 (92 percent) had been approved (VT, 2005). Most applications approved for field testing involved major crops such as corn (over 4,800 applications), soybeans (797), potatoes (745), and cotton (708). Applications approved between 1987 and mid-February 2005 included GE varieties with herbicide tolerance (3,774), insect resistance (3,083), improved product quality (flavor, appearance, or nutrition) (2,241), virus resistance (1,238), agronomic properties like drought resistance (978), and fungal resistance (639).

After extensively field testing a GE variety, an applicant may petition USDA to deregulate (grant permission to produce and sell) the product. If, after extensive review, USDA determines that the new variety poses no significant risk to agriculture or the environment, permission is granted. As of February 2005, USDA had received 103 petitions and granted 63 (including 17 for corn, 11 for tomato, 9 for cotton, 5 for soybeans, and 5 for potatoes). Thirty-six percent of the released varieties have herbicide-tolerance traits, 27 percent have insect-resistance traits, and 17 percent have product-quality traits (VT, 2005).

Extent of Adoption of GE Crops

Driven by farmers'expectations of higher yields, savings in management time, and lower pesticide costs, the rate at which farmers adopt GE crop varieties has risen steadily despite consumer resistance in some countries. An estimated 200 million acres of GE crops with herbicide tolerance and/or insect resistance were cultivated in 17 countries worldwide in 2004, a 20-percent increase over 2003, and U.S. acreage accounts for 59 percent of this amount (Argentina for 20 percent, Canada and Brazil 6 percent each, and China 5 percent) (ISAAA, 2004).

GE varieties of soybeans, corn, and cotton have been available commercially in the U.S. since 1996. Since then, their rate of use by U.S. farmers has climbed most years (Figure 14).

For the most part, farmers have adopted herbicide-tolerant (HT) varieties, which help control weeds, faster than insect-resistant varieties.

Weeds are such a pervasive pest for soybeans, corn, and cotton that over 90 percent of planted acreage for each crop was treated with herbicides in recent years. Acreage share for HT soybeans has expanded more rapidly than that for HT varieties of cotton and corn, reaching 87 percent of U.S. soybean acreage in 2005. Farmers'adoption of HT soybeans has been widespread among major growing States, ranging in 2005 from 76 percent in Michigan to 95 percent in South Dakota. Acreage share for HT cotton has also expanded rapidly, reaching 61 percent in 2005. In contrast, acreage share for HT corn reached only 26 percent in 2005, but this has also trended upward since 2001 (Fernandez-Cornejo, 2004b).

Insect-resistant crops contain a gene from a soil bacterium, *Bacillus thuringiensis* (Bt), which produces a protein toxic to specific insects. Acreage shares for Bt cotton and corn are lower than those for HT soybeans and cotton and vary much more across producing States, with adoption more concentrated in areas with high infestations of targeted pests (insect infestation varies much more widely across locations than does weed infestation). Farmers planted Bt cotton to control tobacco budworm, bollworm, and pink bollworm on 52 percent of cotton acreage in 2005. Acreage share ranged from 13 percent in California to 86 percent in Louisiana. Bt corn, originally developed to control the European corn borer, was planted on 35 percent of corn acreage in 2005, up from 29 percent in 2003 and 24 percent in 2002. The recent increases in acreage share may be largely due to the commercial introduction in 2003/04 of a new Bt corn variety that is resistant to the corn rootworm, a pest that may be even more destructive to corn yields than the European corn borer.

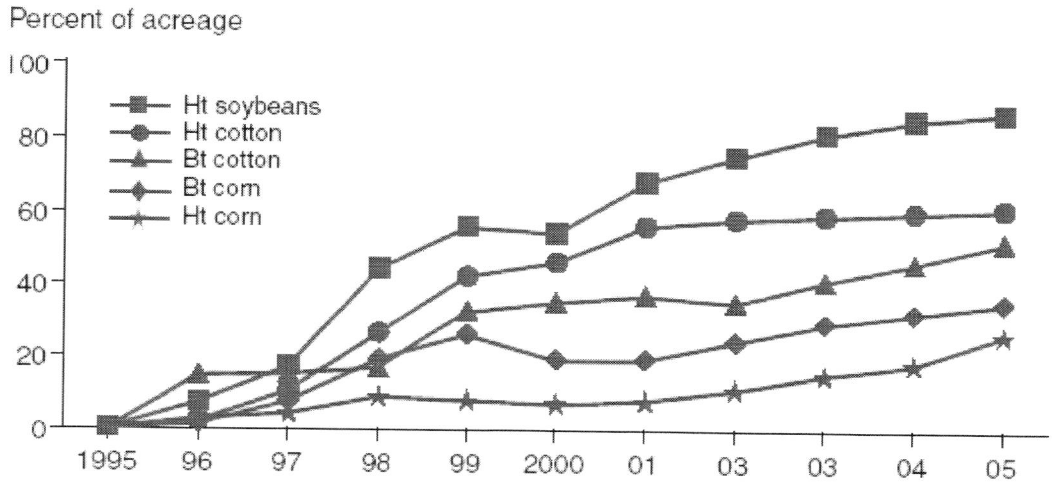

Data for each crop category include varieties with stacked traits.
Source: ERS elaboration from several USDA surveys.

Figure 27. Adoption of genetically engineered crops in the U.S.

Other GE crops used by U.S. farmers over the past 10 years include herbicide-tolerant canola, Bt potatoes (introduced by Monsanto in 1996 and withdrawn from the market after the 2001 season), virus-resistant papaya (developed by Cornell University and University of Hawaii and introduced commercially in 1998), and virus-resistant squash (Table 11). In addition, a tomato genetically engineered to remain on the vine longer and ripen to full flavor after harvest was introduced by Calgene in 1994, but withdrawn after being available sporadically for several years (Colorado State University, 2004).

Main Reasons Stated by U.S.Farmers for Adopting GE Crops

According to surveys conducted by USDA in 2001-03, most farmers (59-79 percent) adopting GE corn, cotton, and soybeans indicated that they did so mainly to "increase yields through improved pest control" (Figure 28). The second most cited aim was to "save management time and make other practices easier" (15 to 26 percent, except for Bt corn, which was much lower); the third reason was to "to decrease pesticide costs" (9-17 percent of adopters). All other reasons combined accounted for 3-7 percent of adopters. Hence, factors expected to increase economic profitability by increasing revenues per acre (yield times price of the crop) or reducing costs (operator labor, pesticides) are expected to promote adoption most.

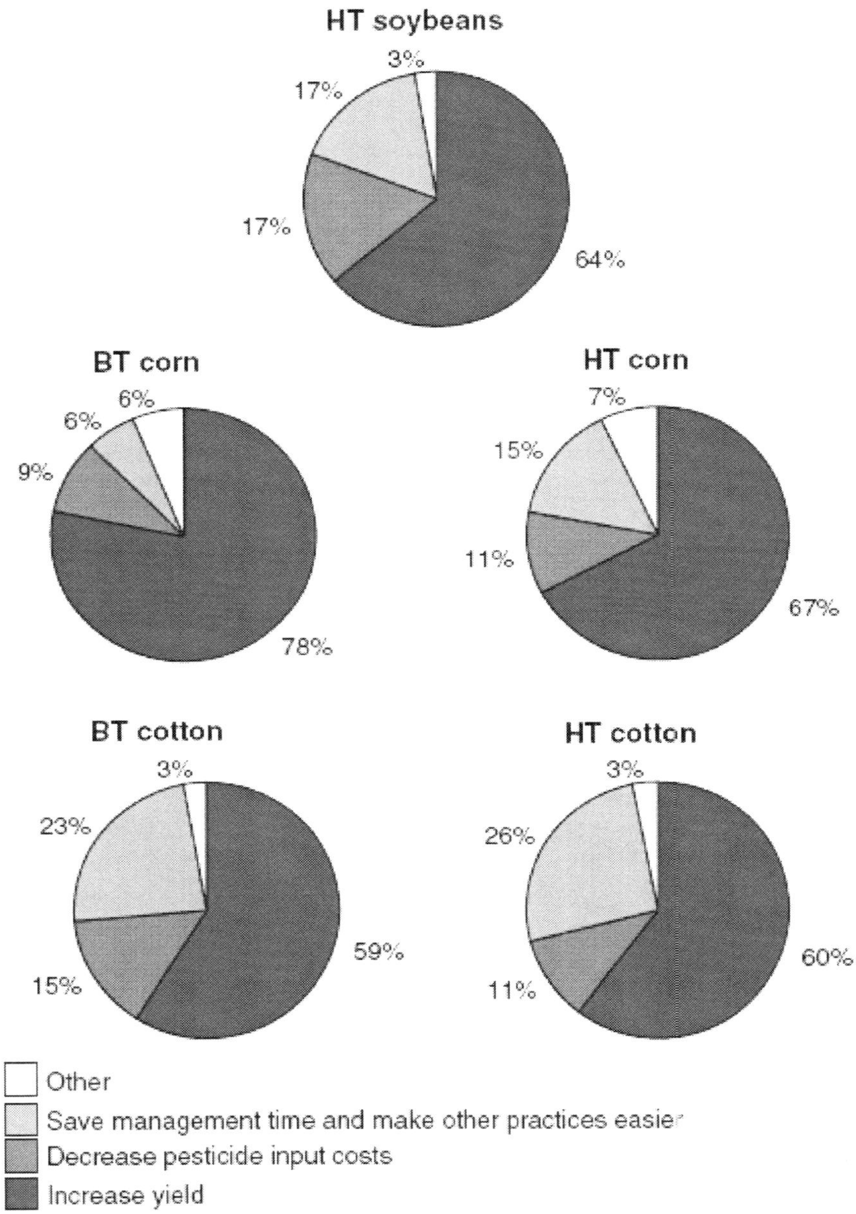

Source: 2004 USDA Agricultural Resource Management Survey, Economic Research Service, USDA.

Figure 28. Main reasons for adopting GE crops, according to farmers.

ADOPTION OF GE CROPS AND YIELDS

The first generation of GE crops does not increase the yield potential of a hybrid. In fact, yield potential may even decrease if the varieties used to carry the herbicide-tolerant or insect-resistant genes are not the highest yielding cultivars. However, by protecting the plant from certain pests, GE crops can prevent yield losses compared with non-GE hybrids, particularly when pest infestation occurs. This effect is particularly important in the case of Bt crops. Before the commercial introduction of Bt corn in 1996, the European corn borer was only partially controlled using chemical insecticides. The economics of chemical use was not always favorable, and timely application was difficult. For these reasons, many farmers accepted yield losses rather than incur the expense of chemical pesticides to treat the insect. Consequently, the use of Bt corn often resulted in yield gains rather than pesticide savings. On the other hand, a different Bt corn trait selected for resistance against the corn rootworm, previously controlled using chemical insecticides, may provide substantial insecticide savings. This new Bt corn variety was recently introduced commercially.

An ERS study estimated the impact of adopting GE crops on yields using an adoption model and 1997 survey data (Fernandez-Cornejo and McBride, 2002, pp. 20-23). The study shows that an increase of 10 percent in the adoption of HT cotton led to a 1.7-percent increase in yields. Similarly, the adoption of Bt cotton in the Southeast was related to a significant increase in yields. On the other hand, the adoption of HT soybeans was related to only small (but still significant) increases in yields.

ADOPTION, NET RETURNS, AND HOUSEHOLD INCOME

According to an ERS study, the impacts of GE crop adoption on U.S. farmers vary by crop and technology (Fernandez-Cornejo and McBride, 2002, pp. 20-25). The main results of the ERS study are presented below.

- ***Planting HT cotton and corn was associated with increased producer net returns, but HT corn acreage was limited.*** The limited acreage on which herbicide-tolerant corn has been used is likely to be acreage with the greatest comparative advantage for this technology. The positive financial impact of adoption may also be due to seed companies' setting low premiums for herbicide-tolerant corn relative to conventional varieties in an attempt to expand market share. Limited adoption of HT corn may be due to constraints imposed on rotation with soybeans. Also, some HT corn varieties have limited approval outside the U.S., restricting their export market potential.
- ***Adoption of Bt cotton and corn was associated with increased returns when pest pressures were high enough.*** The adoption of Bt cotton had a positive association with producer net returns in 1997, but the association was negative for Bt corn in 1998. This suggests that Bt corn may have been used on some acreage where the value of protection against the European corn borer (ECB) was lower than the premium paid for the Bt seed. Because pest infestations differ across the country, the economic benefits of Bt corn are likely to be greatest where target pest pressures are most severe. The decision to use Bt corn must be made before observing the ECB

pest pressure, and damage caused by the ECB varies from year to year. Some farmers may incorrectly forecast infestation levels, corn prices, and yield losses due to infestations, resulting in "overadoption."Also, producers may be willing to pay a premium for Bt corn because it reduces the risk of significant losses if higher-than-expected pest damage does occur.

- ***Despite the rapid adoption of HT soybeans by U.S. farmers, no significant impact on net farm returns was evident in 1997 or 1998.*** This lack of profitability suggests that other factors may be driving adoption for many adopters, such as the simplicity and flexibility (less management time) of weed control. This implies more time available to off-farm employment by farm operators and their spouses. (On average, off-farm earned income is more than twice the net income earned from farming.)
- ***Recent ERS research using 2000 data showed that adoption of HT soybeans was associated with significantly higher off-farm household income for U.S. soybean farmers.*** On-farm household income was not significantly related to adoption, but total farm household income is significantly higher for adopters.

Adoption and Pesticide Use

On the environmental side, pesticide use on corn and soybeans has declined since the introduction of GE corn and soybeans in 1996 (Figure 29).

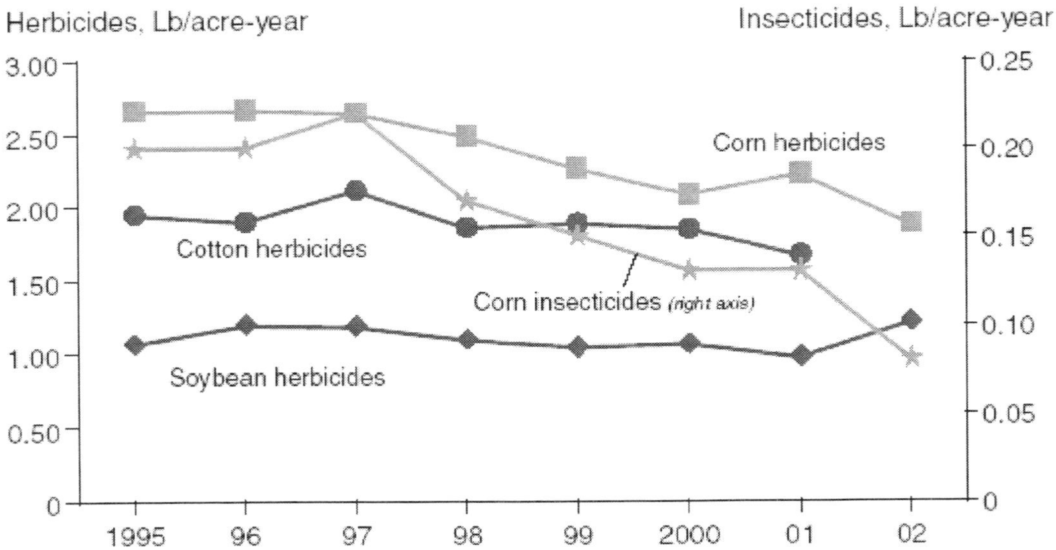

Source: USDA, NASS surveys.

Figure 29. Pesticide use in major field crops.

In addition, ERS research suggests that, controlling for other factors, pesticide use declined with adoption. The overall reduction in pesticide use associated with the increased adoption of GE crops (Bt cotton; and HT corn, cotton, and soybeans, using 1997/1998 data) also resulted in a significant reduction in potential exposure to pesticides. The decline in pesticide applications was estimated to be 19.1 million acre-treatments (Fernandez-Cornejo and McBride, 2002, pp. 26-28). Total pesticides applied to corn, soybeans, and cotton declined by about 2.5 million pounds (active ingredients), despite the (slight) net increase in the amount of herbicides applied to soybeans. For more information on pesticide use, see Chapter 14, "Pest Management".

ADOPTION AND CONSERVATION TILLAGE

The environmental impact of conservation tillage (including no-till, ridge-till, and mulch-till) is well documented. Conservation tillage reduces soil erosion by wind and water, increases water retention, and reduces soil degradation and water/chemical runoff. For more on conservation tillage, see Chapter 13, "Soil Management and Conservation".

According to USDA survey data, the portion of acreage planted with HT soybeans under conservation tillage was larger than the portion of acreage growing conventional soybeans. About 60 percent of the area planted with HT soybeans was under conservation tillage in 1997 (Figure 30), versus 40 percent of conventional soybeans.

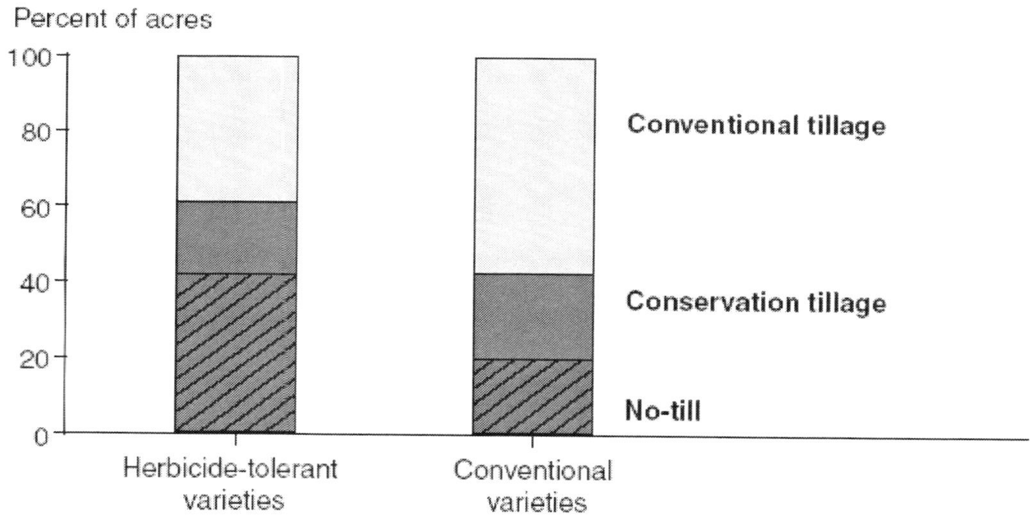

Source: Fernandez-Cornejo and McBride (2002).

Figure 30. Soybeans area under conservation tillage and no-till, 1997.

Differences in the use of no-till between adopters and nonadopters of HT soybeans are even more pronounced: 40 percent versus 20 percent. As a result, adoption of HT crops may indirectly benefit the environment by encouraging the adoption of soil conservation practices that control soil erosion, soil degradation, and runoff.

ECONOMIC BENEFITS OF GE CROPS

GE crops can offer producers distinct advantages over conventional varieties, such as higher yields and lower pest control costs. But producers are not the only ones to gain from the adoption of GE crops. Biotechnology developers and seed companies gain by charging technology fees and seed premiums to adopters of GE varieties. Ultimately, U.S. and foreign consumers may benefit from GE crops through lower commodity prices, which result from increased supplies.

ERS estimated the total market benefit arising from the adoption of three biotech crops in 1997: herbicide-tolerant soybeans, insect-resistant (Bt) cotton, and herbicide-tolerant cotton. Estimated benefits were around $210 million for Bt cotton, $230 million for HT cotton, and $310 million for HT soybeans (Price et al., 2003). This benefit includes the change in total welfare in both the seed input and commodity output markets. Estimated benefits and their distribution depend particularly on the analytical framework, supply and demand elasticity assumptions, crops considered, and year-specific factors (such as weather).

There are tangible benefits to farmers who adopt first-generation GE crops. Not all of the benefits are reflected in standard measures of net returns. As in all studies, results should be interpreted carefully, especially since the impact studies are based on a few years of data. The impacts of GE crops vary with several factors, most notably annual pest infestations, seed premiums, prices of alternative pest control programs, and any premiums paid for segregated (i.e., non-GE) crops. These factors will continue to change over time as technology, marketing strategies for GE versus conventional crops, and consumer perceptions evolve.

REFERENCES

[1] Colorado State University, Department of Soil and Crop Sciences (2004). "Discontinued Transgenic Products," *Transgenic Crops: An Introduction and Resource Guide.*

[2] Comis, D. (1997). *Safe corn pest bait expected to slash U.S. insecticide use.* U.S. Department of Agriculture, Agricultural Research Service.

[3] Fernandez-Cornejo, J. (2004a). *The Seed Industry in U.S. Agriculture: An Exploration of Data and Information on Crop Seed Markets, Regulation, Industry Structure, and Research and Development.* AIB-786. U.S. Dept. Agr., Econ. Res. Serv., Feb.

[4] Fernandez-Cornejo, J. (2004b). "Adoption of Genetically Engineered Crops in the U.S." Data product. U.S. Dept. Agr., Econ. Res. Serv.

[5] Fernandez-Cornejo, J., and W.D. McBride (2002). *The Adoption of Bioengineered Crops.* AER-810. U.S. Dept. Agr., Econ. Res. Serv., May.

[6] Hernandez-Cornejo, J., M. Caswell, and C. Klotz-Ingram (1999). "Seeds of Change: From Hybrids to Genetically Modified Foods," *Choices* (Millennium issue, 4th quarter): pp. 18-22.

[7] International Service for the Acquisition of Agri-biotech Applications (ISAAA) (2004). *Preview: Global Status of Commercialized Biotech/GM Crops:* ISAAA Briefs 32-2004.

[8] NRC (National Research Council) (2002). *Animal Biotechnology: Science Based Concerns.* Board on Agriculture and Natural Resources, Board on Life Sciences, The National Academies Press. Washington, DC.

[9] Panos (1998)."Greed or Need? Genetically Modified Crops." Panos Media Briefing No. 30, Oct.

[10] Pew Initiative on Food and Biotechnology (2001). *Harvest on the Horizon: Future Uses of Agricultural Biotechnology.* Available at http://pewagbiotech.org/research/harvest/

[11] Price, G.K., W. Lin, and J. Fernandez-Cornejo (2003). *The Size and Distribution of Market Benefits from Adopting Agricultural Biotechnology.* TB-1906. U.S. Dept. Agr., Econ. Res. Serv., Nov.

[12] Shoemaker, R., J. Harwood, K. Day-Rubenstein, T. Dunahay, P. Heisey, L. Hoffman, C. Klotz-Ingram, W. Lin, L. Mitchell, W. McBride, and J. Fernandez-Cornejo (2001). *Economic Issues in Agricultural Biotechnology.* AIB-762, U.S. Dept. Agr., Econ. Res. Serv., Feb.

[13] UC Vegetable Research and Information Center, Pest-Resistance Management (2005). "Insecticidal Bt-Potatoes." University of California.

[14] U.S. Department of Agriculture (2004). "Agricultural Biotechnology, Frequently Asked Questions: What Is Biotechnology?"

[15] Virginia Polytechnic Institute and State University (VT) (2005). "Information Systems for Biotechnology: Field Test Releases in the U.S."

In: Agricultural Resources and Environmental Indicators ISBN 978-1-60021-467-7
Editor: Keith Wiebe and Noel Gollehon, pp. 91-96

Chapter 10

PRODUCTIVITY AND OUTPUT GROWTH IN U.S. AGRICULTURE

Eldon Ball

U.S. agricultural output grew at an average annual rate of 1.76 percent over 1948-2002. Input use actually declined in aggregate, so the positive growth in farm sector output was wholly due to productivity growth.

INTRODUCTION

U.S. agricultural output has more than doubled in the last 50 years, growing at an average rate of 1.76 percent per year (Figure 31). This rate is a remarkable achievement considering that labor has been departing the sector and land use has declined slightly, while capital influx has been modest. In spite of the growth in materials like fertilizer, fuel, and machinery, the net contribution of all inputs was slightly negative, leaving productivity growth as the sole source of output growth. While the contribution of other factors like labor, capital, and production inputs has risen or fallen with macroeconomic trends, one intangible input—productivity—has grown inexorably. But what is productivity?

Productivity is not equivalent to output (or production). Productivity reflects improvements in the ability to transform inputs into outputs. In the most literal sense, it is a residual measure of the contribution to output growth after all other factors have been accounted for. It is the nonphysical product of innovation, efficiency, management, research, weather, and luck. And its rate of growth seems to have slowed in recent years, coincident with a dropoff in public funding for agricultural research since the 1980s.

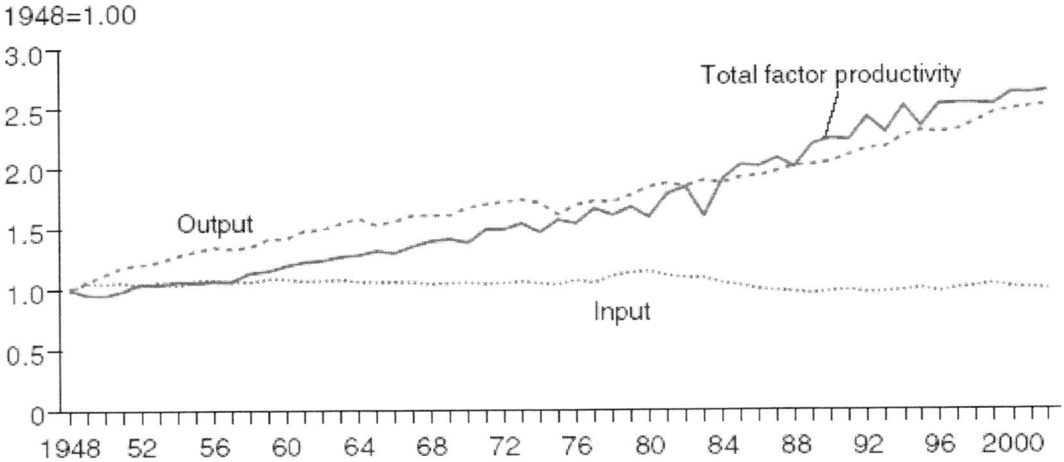

Source: ERS-USDA, from information in the "Agricultural Productivity in the United States" data product.

Figure 31. U.S. agricultural output, input, and total factor productivity, 1948-2002.

What Is Productivity Growth?

Productivity growth is a reflection of technological change and efficiency improvements, e.g., better management or economies of scale, which result in producing more output from a given level of input. Productivity growth is difficult to measure. Measuring productivity growth requires the careful accounting for all outputs and inputs and especially attempting to measure the improvements and actual flow of services from inputs, i.e., what is actually used. Once all inputs, including their technical improvements, are measured, what is not captured—the residual—is called productivity growth.

Patterns in Output and Productivity Growth

Output growth derives from growth in the use of inputs (capital, land, labor, materials) and total factor productivity growth. Input growth has been the main source of economic growth for the U.S. economy as a whole and for most sectors. Only in agriculture does productivity growth exceed input growth (Table 15), over 1948-2002 and in 10 subperiods.

Labor

The singular importance of the role of productivity growth in agriculture is all the more remarkable given labor's long-term contraction. Over 1948-2002, labor input declined, on average, 2.4 percent each year, a rate unmatched by any non-farm sector. The historic decline in farm labor—both farmers and farm laborers—occurred as workers sought higher wages and other income opportunities in the non-farm sector. This rate of decline in labor appears to have slowed since the 1980s (Figure 32) as average household incomes in the farm and non-farm sectors have converged (Hoppe) Farm households, like non-farm households, now pursue multiple careers and diversify their earnings. In fact, the income available to the average farm household can support a standard of living equal to or above that of the average non-farm household, reducing the desire to leave farming.

Capital

Capital input in agriculture exhibits a different pattern than labor. During 1973-79, U.S. agriculture experienced rapid growth, fueled by a growth in exports resulting from increased global liquidity, rising incomes, and production shortfalls in other parts of the world. U.S. farm exports surged from an average $4.8 billion in 1950-70 to $9.4 billion in 1972 and $17.7 billion in 1973. Exports continued to increase through 1981, when they peaked at $43.3 billion. In addition, domestic forces—including a drop in interest rates and rising inflation—contributed to an increase in borrowing for the purchase of land and equipment. For much of the 1970s, real interest rates were close to zero and at times negative, reducing the cost of capital. Capital input in agriculture increased 2 percent per year between 1973 and 1979, adding an average 0.33 percentage points per year to output growth (Table 15).

Table 15. Source of growth for U.S. farm sector

Item	1948-2002	1948-1953	1953-1957	1951-1960	1960-1966	1966-1969	1969-1973	1973-1979	1979-1989	1989-1999	1999-2002
					Percent growth per year						
Output growth	1.76	1.76	0.89	4.44	1.06	2.26	2.51	2.53	1.00.	2.21	-0.40
Sources of output growth											
Labor	-0.62	-1.16	-1.05	-0.75	-1.10	-1.05	-0.28	-0.72	-0.41	-0.09	-0.44
Capital	0.01	0.69	0.15	0.07	0.11	0.35	0.17	0.33	-0.55	-0.23	-0.03
Land	-0.06	0.02	-0.11	-0.10	-0.05	-0.15	-0.22	0.00	-0.08	0.00	-0.06
Materials	0.64	1.54	1.10	1.50	0.69	0.30	0.64	1.50	-0.64	1.13	-1.26
Total factor productivity	1.79	0.65	0.81	3.73	1.40	2.81	2.221	1.40	2.69	1.41	1.38

Source: ERS-USDA from information in the data product, "Agricultural Productivity in the United States" on the ERS website.

However, the economic environment changed in the early 1980s. A change to restrictive monetary policy by the Federal Reserve pushed interest rates up sharply. The dollar appreciated on foreign exchange markets, and world export prices fell. The average real interest cost on variable-rate debt rose to nearly 16 percent in 1981-83. Real interest rates remained high thereafter, as the stringency of Federal Reserve policy was heightened due to large fiscal deficits. This mix of fiscal stimulus and monetary restraint slowed the growth in export-dependent sectors of the economy, including agriculture. The value of U.S. farm exports fell from $43.3 billion in 1981 to $26.2 billion in 1986, as both volume and prices dropped. Growth in agricultural output slowed to about 1 percent per year during 1979-89, versus 2.5 percent over 1973-79 (Table 15). Capital's contribution to output growth was negative during this period, averaging –0.55 percentage points per year.

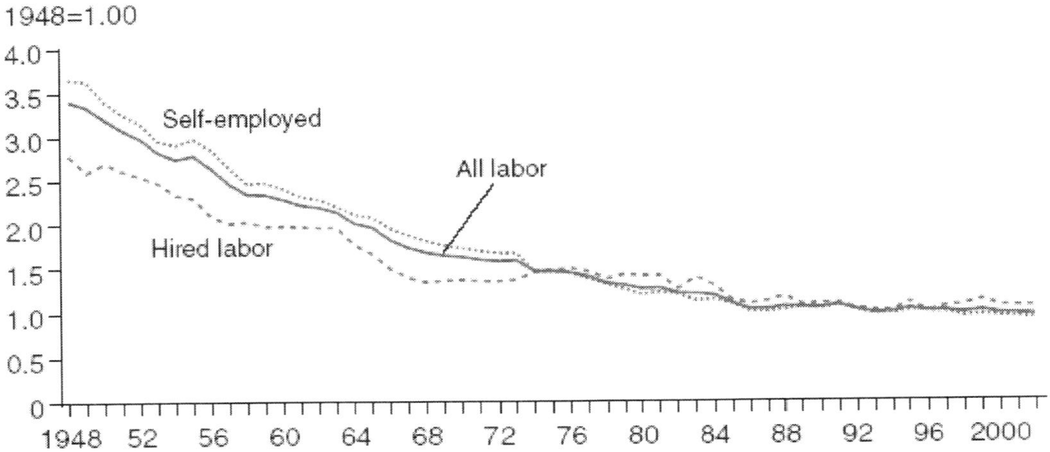

Source: ERS-USDA, from information in the "Agricultural Productivity in the United States" data product on the ERS website.

Figure 32. The secular decline in agricultural labor has slowed down, 1948-2002.

Land and Material Inputs

Land's contribution to growth in agricultural output was negative for all recent time periods but 1948-53, 1973-79, and 1989-99. Over 1948-2002, the contribution of land to output growth was –0.06 percentage points per year. It seems ironic that the contribution of land to output growth would generally be negative in a land-based industry like agriculture. The explanation lies in the vast availability of farmland in the United States. The positive growth in materials reflects the substitution of those inputs for land. Material inputs'contribution averaged 0.64 percent per year over 1948-2002. Still, this did not offset the negative contributions of labor and land, making the contribution of all inputs negative

Parallels can be drawn between the 1973-79 and 1989-99 periods. Both were periods of rapid output growth, fueled largely by growth in demand for agricultural exports. And input growth accounted for a disproportionate share of output growth during both periods. Growth in intermediate inputs contributed more than 1 percentage point per year to output growth during 1989-99. The net contribution of input growth to output growth was 0.8 percentage point per year during 1989-99, versus 1.02 percentage points during 1973-79.

How Is Productivity Measured?

The U.S. Department of Agriculture (USDA) has been monitoring agriculture's productivity performance for decades. In fact, USDA was the first Federal agency in 1960 to introduce multifactor productivity measurement into the Federal statistical program. Today, ERS routinely publishes total factor productivity (TFP) measures from production accounts that distinguish multiple outputs and inputs and adjusts for quality change in each input category. Its TFP model is based on the translog transformation frontier. It relates the growth of multiple outputs to the growth rates of capital, land, labor, and intermediate inputs, weighted by their shares in total costs. The changing demographic character of the agricultural workforce is used to build a quality-adjusted index of labor input. Similarly, much asset-specific detail underlies the measure of capital input. The contribution of feed and seed, chemicals, and energy are captured in the index of intermediate inputs. An important innovation is the use of hedonic price indexes in constructing measures of fertilizer and pesticide consumption. The result is a series of TFP indexes spanning 1948 to 2002.

ERS defines the farm sector as it is defined in the U.S. national income and product accounts. Production of goods and services that are secondary to agriculture is assigned to the primary producing industry. This enables certain secondary activities closely linked to agriculture for which information on production and input use cannot be separately observed to be included in the total factor productive activity of agriculture. Examples include the provision of machine services, contract feeding of livestock, recreational activities, and other activities involving the use of the land and the means of agricultural production.

Productivity Growth

Since productivity grew 1.79 percent per year over the period of 1948-2002, farm sector productivity in 2002 was 263 percent above its 1948 level. As a consequence, and in the absence of input growth between 1948 and 2002,

Productivity growth single-handedly caused farm output to grow 259 percent above its 1948 level.

Looking at productivity trends over the long term is appropriate. Productivity is largely the result of long-term investments in scientific research, so while agricultural productivity has risen and fallen year to year—typically driven by year-to-year fluctuation in output due to weather—it has generally trended upward. However, since 1996, productivity growth has slowed. Is this a change in trend? A key source of productivity growth—public investments in research—has been flat in real terms since the 1980s. (See Chapter 8, Agricultural Research and Development".) Only time will tell how this may affect future productivity growth.

REFERENCES

[1] Ball, V.E., J.C. Bureau, R. Nehring, and A. Somwaru (1997). "Agricultural Productivity Revisited," *American Journal of Agricultural Economics 79: 1045-1063.*

[2] Ball, V.E., F. Gollop, A. Kelly-Hawke, and G. Swinand (1999). "Patterns of State Productivity Growth in the U.S. Farm Sector: Linking State and Aggregate Models," *American Journal of Agricultural Economics* 81:164-179.

[3] Hoppe, Bob. "Farm Structure: Farm Operator Household Income Data." in the "Farm Household Economics and Well-being." Briefing Room. U.S. Dept. Agr., Econ. Res. Serv.

[4] Mishra, A.K., H.S. El-Osta, M.J. Morehart, J.D. Johnson, and J.W. Hopkins (2002). *Income, Wealth, and the Economic Well-Being of Farm Households.* AER-812 U.S. Dept. Agr., Econ. Res. Serv.

[5] Jorgenson, D., and F. Gollop (1992). "Productivity Growth in U.S. Agriculture: A Postwar Perspective," *American Journal of Agricultural Economics* 74: 745-50.

[6] Jorgenson, D., F. Gollop, and B. Fraumeni (1987). *Productivity and U.S. Economic Growth.* Cambridge, MA: Harvard University Press.

[7] Kendrick, J., and E. Grossman (1980). *Productivity in the United States, Trends and Cycles.* Baltimore, MD: The Johns Hopkins University Press.

In: Agricultural Resources and Environmental Indicators ISBN 978-1-60021-467-7
Editor: Keith Wiebe and Noel Gollehon, pp. 97-105

Chapter 11

GLOBAL RESOURCES AND PRODUCTIVITY

Keith Wiebe

Global food production has grown faster than population in recent decades, due largely to improved seeds and increased use of fertilizer and irrigation. Soil degradation, which depends on farmers'incentives to adopt conservation practices, has slowed yield growth in some areas but does not threaten food security at the global level.

INTRODUCTION

Increased resource use and improvements in technology and efficiency have increased global food production more rapidly than population in recent decades, but 800 million people remain food insecure (Figure 33).

Meanwhile, growth in global agricultural productivity appears to be slowing, and land degradation has been blamed as a contributing factor. The interactions between biophysical processes and economic choices are complex, and data necessary to measure these processes are scarce, so estimates of land degradation's impact on productivity vary widely—as high as 8 percent per year due to soil erosion alone in the United States and as low as 0.1 percent per year due to all forms of soil degradation on a global scale. These differences make it difficult to assess potential impacts on food security or the environment, and thus the appropriate nature and magnitude of policy response.

Improvements in economic analysis of geographic data offer new insights. ERS recently studied how agricultural productivity varies with differences and changes in land quality, and how degradation-induced changes in productivity affect food security. Results indicate that land degradation does not threaten productivity growth and food security at the global level. But problems do exist in some areas, especially where resources are fragile and markets function poorly.

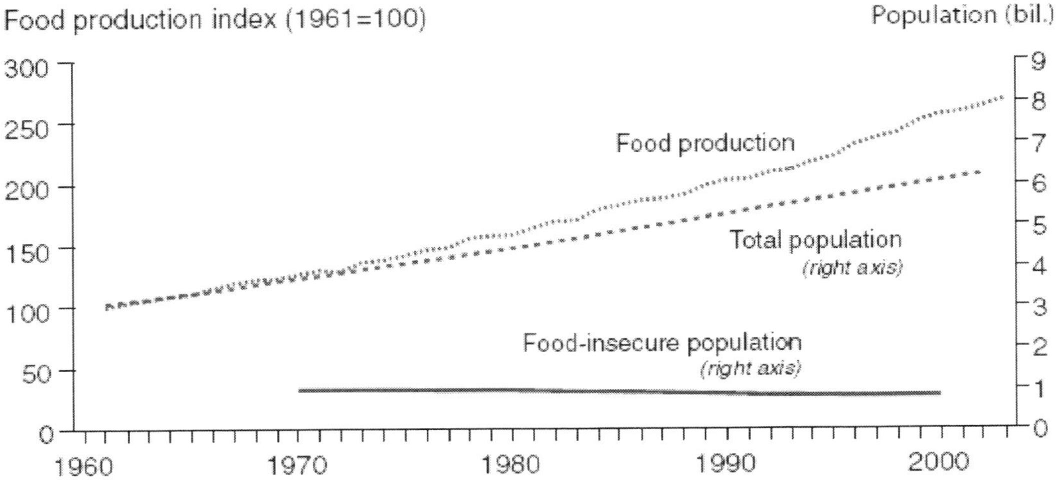

Source: Food and Agricultural Organization.

Figure 33. Food production has outpaced population.

WORLD FOOD SUPPLIES HAVE INCREASED FASTER THAN DEMAND… SO FAR

Global demand for food has increased rapidly since the mid-20th century as a result of growth in population, income, and other factors. The world's population has doubled over the past four decades, to 6.4 billion in 2004. World population growth has slowed in recent years, but is still projected to reach 9 billion by about 2050. Per capita income is projected to grow by an average of about 2 percent per year over the next decade, continuing recent trends. Based on these factors, the Food and Agriculture Organization of the United Nations (FAO) and the International Food Policy Research Institute (IFPRI) project that global demand for cereals will increase by 1.2-1.3 percent per year over the next several decades, while demand for meat will increase slightly faster. Most of the increased demand is projected to come from developing countries, especially from Asia.

Between 1961 and 1999, the FAO's aggregate crop production index grew at an average annual rate of 2.3 percent. Crop production per capita has increased for the world as a whole (at an average rate of 0.6 percent per year), and in all regions except Africa. Global cereal production per capita (Figure 34) has fallen since 1984, with steady increases in Asia offset by long-term declines in sub-Saharan Africa and more recent declines in North America, Europe, Oceania, and the former Soviet Union.

But these more recent declines were due not to binding resource and technology constraints but rather to the combined effects of weak grain prices, policy reforms, and institutional change. (See box for, Definitions)

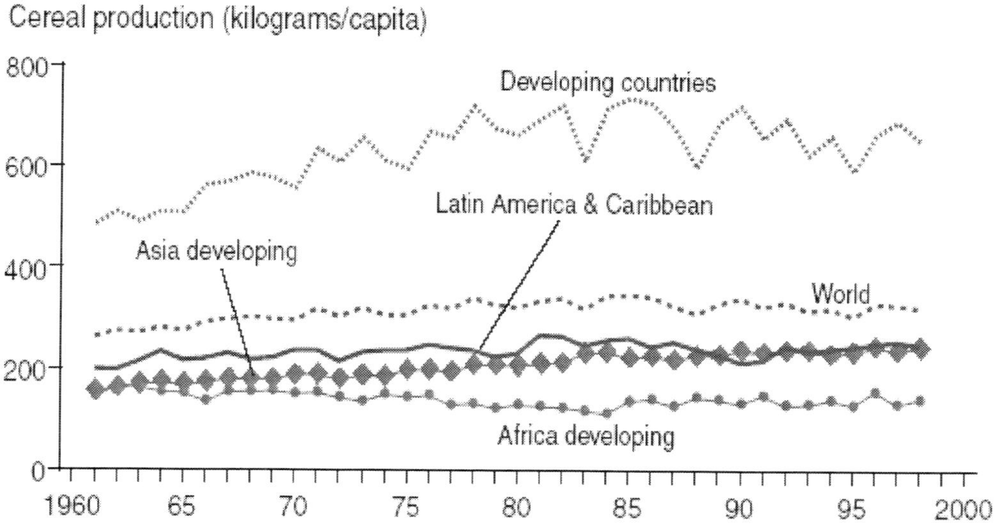

Source: FAO.

Figure 34. Cereal production per capita by region.

Definitions

Land quality—The ability of land to produce goods and services that are valued by humans. This ability derives from inherent/natural attributes of soils (e.g., depth and fertility), water, climate, topography, vegetation, and hydrology, as well as "produced" attributes such as infrastructure (e.g., irrigation) and proximity to population centers.

Land degradation—Changes in the quality of soil, water, and other characteristics that reduce the ability of land to produce goods and services that are valued by humans. Some forms of land degradation, such as nutrient depletion, can be halted and even reversed relatively easily—for example, by balancing nutrient application with that taken up in harvested crops. Other forms of land degradation, such as erosion or salinization, can be slowed or halted through appropriate management practices, but are generally very costly to reverse.

Agricultural productivity—A measure of the amount of agricultural output that can be produced with a given level of inputs. Agricultural productivity can be defined and measured in a variety of ways, including the amount of a single output per unit of a single input (e.g., tons of wheat per acre or per worker), or in terms of an index of multiple outputs relative to an index of multiple inputs (e.g., the value of all farm outputs divided by the value of all farm inputs) (see Chapter 3.4, "Productivity and Output Growth in U.S. Agriculture").

IFPRI projects that world cereal production will increase by about 1.3 percent per year through 2020, enough to raise per capita cereal production by about 0.2 percent annually. Such increases have the potential to satisfy projected food demands (and nutritional requirements) for the foreseeable future, but actual patterns will depend on the availability and quality of productive resources, as well as market incentives, policy measures, and research investments.

Area Growth Is Slowing, So Yields Will Become More Important

FAO reports that the total area devoted to crops worldwide has increased by about 0.3 percent per year since 1961, to 3.8 billion acres in 2002. Growth has slowed markedly in the past decade, to about 0.1 percent per year, as a result of weak grain prices, deliberate policy reforms (in North America and Europe), and institutional change (in the former Soviet Union). FAO estimates that an additional 6.7 billion acres currently in other uses are suitable for crop production, but this land is unevenly distributed, and includes land with relatively low yield potential and significant environmental value.

Given economic and environmental constraints on cropland expansion, the bulk of increased crop production will need to come from increased yields on existing cropland. FAO data indicate that world cereal yields rose by about 2.5 percent per year from 1961 to 1990, but growth slowed to 1.1 percent per year in the 1990s (Figure 35).

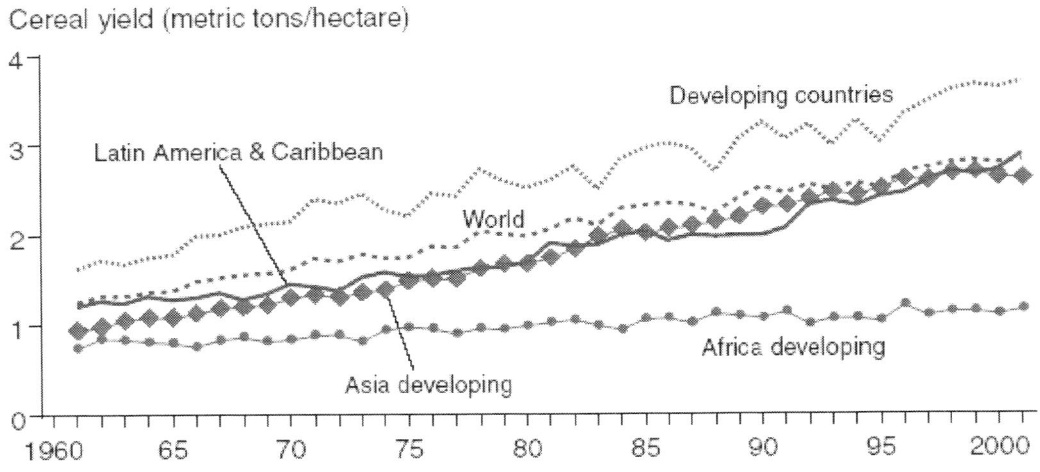

Source: Food and Agricultural Organization.

Figure 35. Cereal yields by region.

As a result of reduced input use (reflecting low cereal prices), market and infrastructure constraints, and low levels of investment in agricultural research and technology, IFPRI and FAO project that yield growth will slow further to about 0.8 percent per year over the next several decades (see Chapter 10, "Productivity and Output Growth in U.S. Agriculture").

Genetic improvements have contributed greatly to gains in yields and production of major crops, beginning with wheat, rice, and maize in the 1960s. About half of all recent gains in crop yields are attributable to genetic improvements. By the 1990s, 90 percent of wheat acreage in developing countries was in scientifically bred varieties, as was 74 percent of land in rice and 62 percent of land in maize. In developed countries, 100 percent of land in wheat, maize, and rice was in scientifically bred varieties by the 1990s (and probably even earlier). Gains from genetic improvements will continue, but likely at slower rates and increasing costs, as gains in input responsiveness have already been largely exploited (see Chapter 7, "Crop Genetic Resources").

FAO data indicate that increased fertilizer consumption accounted for one-third of the growth in world cereal production in the 1970s and 1980s. Growth in fertilizer consumption per hectare of cropland has been slowing, however, from a global average annual increase of about 9 percent in the 1960s to an average annual decline of about 0.1 percent in the 1990s. Among developing regions, per-hectare fertilizer consumption increased most rapidly in land-scarce Asia and most slowly in Africa. Growth in fertilizer consumption also slowed (and even declined) in developed regions, but remains at relatively high levels. Future fertilizer use will need to balance its potential to mitigate onsite land degradation (soil fertility depletion) with the risk of increased offsite degradation (impacts on water quality, for example) (see Chapter 15, "Nutrient Management").

Water will be a critical factor limiting crop production in the 21st century. Agriculture accounts for more than 70 percent of water withdrawals worldwide, and over 90 percent of withdrawals in low-income developing countries. The total extent of irrigated cropland worldwide has grown at an average annual rate of 1.9 percent since 1961, although this rate has been declining. About 18 percent of total cropland area is now irrigated, most of it in Asia. Population growth and the increasing cost of developing new sources of water will place increasing pressure on world water supplies in the coming decades. Even as demand for irrigation water increases, farmers face growing competition for water from urban and industrial users, and to protect ecological functions. In addition, waterlogging and salinization of irrigated land threaten future crop yields in some areas (see AREI Parts 2 and 4).

The Intergovernmental Panel on Climate Change (IPCC), representing a broad scientific consensus, projects that the Earth's climate will change significantly over the course of the 21st century because of increasing concentrations of carbon dioxide and other "greenhouse" gases in the atmosphere. Global crop production would be little affected in aggregate, but potential impacts and adjustment costs vary widely, and could be quite high in some areas. For example, changing patterns of precipitation, temperature, and length of growing season resulting from a doubling of atmospheric concentrations of carbon dioxide would tend to increase agricultural production in temperate latitudes and decrease it in the tropics.

ERS recently examined regional differences in cropland quality using geographic data on land cover, soil, and climate. About 13 percent of global land area has soils and climate that are of high quality for agricultural production (Figure 36).

Land quality changes over time as a result of natural and human-induced processes, but data on these changes are extremely limited. Only one global assessment has been done to date: the Global Land Assessment of Degradation (GLASOD) in 1991 (Oldeman et al., 1991). Based on the judgment of over 250 experts around the world, GLASOD estimated that 38 percent of the world's cropland had been degraded to some extent as a result of human activity since World War II. GLASOD identified erosion as the main cause of degradation (affecting 4 billion acres, mostly in Asia and Africa), followed by loss of soil nutrients (336 million acres, mostly in South America and Africa) and salinization (190 million acres, mostly in Asia).

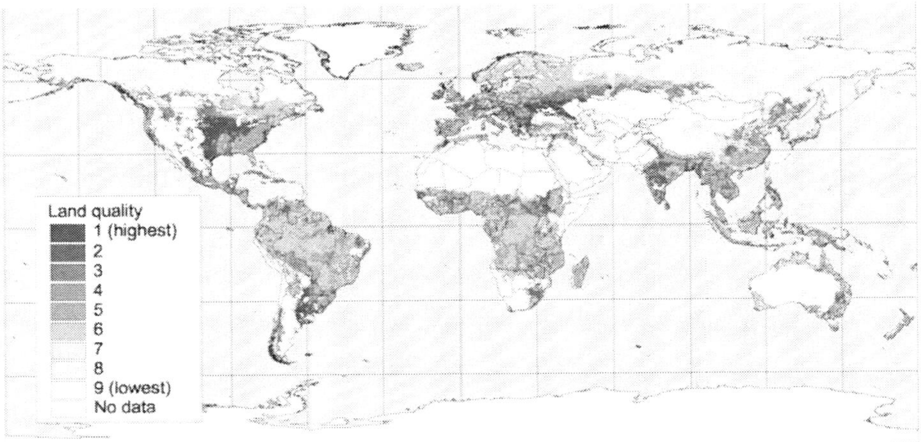

Source: USDA Natural Resources Conservation Service, World Soil Resources Office.

Figure 36. Land quality classes.

Previous studies have sought to measure land quality's role in explaining differences in agricultural productivity between countries, but have only considered factors such as climate and irrigation because of data constraints. ERS researchers incorporated the role of soil characteristics as well, and found that the quality of labor, institutions, and infrastructure also affect productivity. Holding other factors constant, ERS found that the productivity of agricultural labor is generally 20-30 percent higher in countries with good soils and climate than it is in countries with poor soils and climate.

Based on climate and inherent soil properties, scientists with USDA's Natural Resources Conservation Service have estimated water-induced erosion rates that vary widely by crop production area, soil, and region, but range in most cases between 5 and 7 tons per acre per year. Den Biggelaar et al. (2004) recently reviewed over 300 plot-level experiments on yield losses due to soil erosion from around the world and found that for most crops, soils, and regions, yields decline by 0.01-0.04 percent per ton of soil loss. Combining these erosion rates and yield impacts allows estimates of potential annual yield losses to erosion in the absence of changes in farming practices. These estimates vary widely by crop and region. For example, corn yield losses to soil erosion range from an average of 0.2 percent per year in North America to 0.9 percent per year in Latin America. Differences in crop coverage limit comparison of regional totals, but aggregating across regions and crops generates an estimated potential erosion-induced loss of 0.3 percent per year in the value of crop production.

These estimates represent potential impacts of water-induced erosion for selected crops on soils and in regions for which plot-level data were available. Estimated impacts would likely be larger if other degradation processes and crops were considered. On the other hand, actual impacts may also be smaller for any given crop and degradation process to the extent that farmers take steps to avoid, reduce, or reverse land degradation and its impacts.

Farmers have Incentives to Address Land Degradation

Farmers choose between alternative technologies based on biophysical characteristics such as soil quality and access to water, as well as social and economic characteristics that include land tenure, income and wealth, and access to credit and information (see Chapter 12, "Farm Business Management"). Understanding of farmers'incentives is thus critical. For example, practices generating high net returns today may not do so indefinitely if they result in land degradation over time. But practices that reduce land degradation and offer higher net returns over time may require initial investments that inhibit adoption in the short term. ERS researchers explored such tradeoffs in a dynamic analysis of soils and economic data from the north-central United States. Results suggest that actual yield losses under practices that maximize net returns over the long run will typically be lower than potential losses derived from agronomic studies, and are generally less than 0.1 percent per year in the north-central United States.

In order to benefit from a conservation practice that requires an initial investment, a farmer must anticipate farming a particular plot of land long enough to realize the benefit. A farmer with a lease that expires after 1 year, for example, receives only a fraction of the benefit that would be realized by a farmer with a 5-year lease, and both receive less benefit than would a farmer who owns his or her land. ERS research confirms that conservation choices by U.S. corn producers vary significantly with land tenure and the timing of costs and returns to different practices.

Even with secure tenure and the prospect of long-term gains, a farmer might still be unable to afford the initial investment needed to adopt a particular conservation practice, perhaps due to poverty or constraints on access to credit. A farmer might also lack the information needed to compare longrun costs and benefits of alternative practices. Under such market imperfections, optimal choices by farmers would likely result in yield losses greater than those estimated under well-functioning markets, but still less than losses with no farmer response (Figure 37).

Farmers' responses to economic incentives lend support to the lower range of previous estimates of yield losses to land degradation. This does not mean that such losses are unimportant – just that they have historically been masked by increases in input use and improvements in technology and efficiency. Problems do exist in some areas, especially where resources are fragile and markets function poorly. Given projections that yield growth is slowing, yield losses to land degradation are likely to become more of a concern in the future.

Policymakers Play a Critical Role in Shaping Farmers' Incentives

When markets function well, private incentives to reduce land degradation will likely suffice to address on-farm productivity losses. When markets function poorly, private incentives are diminished. Policymakers play a critical role in establishing and maintaining the physical and institutional infrastructure necessary to allow markets to function effectively. This includes transportation and communication networks that facilitate input and output markets, as well as stable and transparent legal and political institutions that encourage longer-term planning horizons. Clear and enforceable property rights are critical in providing incentives for landowners to conserve or enhance land quality.

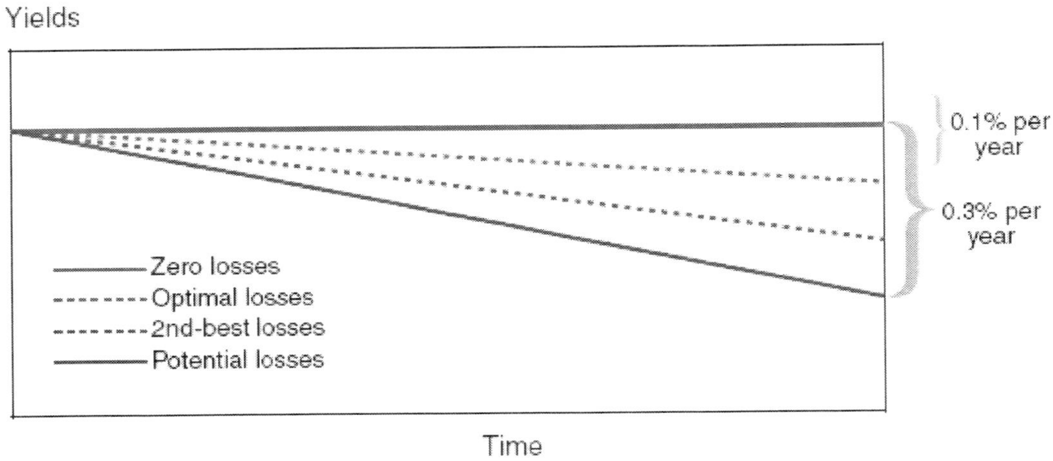

Source: Economic Research Service, USDA.

Figure 37. Yields at different rates of land degradation.

In some circumstances, it may also be necessary to offer direct payments to enhance farmers'incentives to adopt conservation practices. Such payments are well-established in conservation programs in the United States and in many other countries, but require careful attention to the timing and magnitude of payments in order to sustain incentives (see Chapter 21, "Conservation Policy Overview"). While such approaches pose daunting challenges in terms of implementation, they may also help achieve the broader agricultural, environmental, and food security objectives of the World Food Summit, the United Nations Convention to Combat Desertification, and other multilateral initiatives.

References

[1] den Biggelaar, Christoffel, Rattan Lal, Keith Wiebe, Hari Eswaran, Vince Breneman, and Paul Reich (2004). "The Global Impact of Soil Erosion on Productivity," *Advances in Agronomy* 81: 1–95.

[2] Eswaran, Hari, and Paul Reich (2001). World Soil Resources Map Index, U.S. Dept. Agr., Natural Resources Conservation Service.

[3] Oldeman, L.R., R.T.A. Hakkeling, and W.G. Sombroek (1991). "World Map of the Status of Human-Induced Soil Degradation: A Brief Explanatory Note." International Soil Reference and Information Centre and United Nations Environment Programme.

[4] Shapouri, Shahla, and Stacey Rosen (2000). *Food Security Assessment.* International Agriculture and Trade Report, GFA-12, U.S. Dept. Agr., Econ. Res. Serv., Dec.

[5] Wiebe, Keith (2003). *Linking Land Quality, Agricultural Productivity, and Food Security.* AER-823, U.S. Dept. Agr., Econ. Res. Serv.

[6] Wiebe, Keith (ed.) (2003). *Land Quality, Agricultural Productivity, and Food Security: Biophysical Processes and Economic Choices at Local, Regional, and Global Scales.* Cheltenham (UK) and Northampton, MA (U.S.): Edward Elgar Publishing Co.

PART 4: AGRICULTURAL PRODUCTION MANAGEMENT

In: Agricultural Resources and Environmental Indicators ISBN 978-1-60021-467-7
Editor: Keith Wiebe and Noel Gollehon, pp. 109-116

Chapter 12

FARM BUSINESS MANAGEMENT

Jim Johnson and Mitchell Morehart

Assessments of agricultural resources typically focus on natural resources (like land and water) and physical inputs (like fertilizer and pesticides). Assessments of farm diversity commonly focus on aspects of farm structure (like output and organization) and farm household characteristics (like farm and off-farm sources of income and wealth). Relatively unexplored are farms'management resources, which are a key input into agricultural production in their own right and also help shape decisions about the availability and use of other resources, including land and water. This chapter examines the wide range of stakeholders who control farm resources and influence management decisions about agricultural resource use.

INTRODUCTION

This chapter examines the management structure of farms to ascertain who controls the use of farm assets, including land and water. Management units that make decisions for farms are described, extending information about how farmers control and guide their businesses. The chapter also examines decisions of farmers from the perspective of how production, marketing, finance, and human resources are used to form farm businesses.

CHARACTERISTICS OF FARM BUSINESSES' MANAGERS

A farm's management unit consists of the individual or group responsible for decisions about how a farm will be operated. How a farm is legally organized is often viewed as being the same as its management. A proprietor makes decisions for proprietorships, partners for partnerships, and elected directors and officers for corporate farms. However, a farm's management unit may not be synonymous with its ownership. For example, land owners may or may not participate in management decisions. The Census of Agriculture reported in 1999 that 14 percent of landlords either made or shared in decisions related to selection of fertilizer and chemicals, while 13 percent helped decide cultivation practices (USDA, 1999).

Legal organization, while helpful in indicating a farm's governance structure, may not reveal who participates in farm management. Even on proprietor farms, more than one person may participate in management decisions (see box, "Farm Business Management"). Of the 2.1 million U.S. farms reported by the Census of Agriculture, over 121,000 reported three or more operators (USDA, 2004). Together, farms reported 2.7 million operators. In 2003, the primary operator made crop decisions on 52 percent of farms, two operators made joint decisions on 12 percent of farms, and a third operator was involved on 1.3 percent of farms (Figure 38). For the remaining 33 percent of farms, crop production was likely not a part of production activities. In the last decade, involvement by persons other than the primary operator has remained an important aspect of farm management.

Farm Business Management

Participants in management decisions

Principal farm operators Family members Other operators	Landlords Hired advisors Informal advisors Hired farm managers

↓

Farm management decisions

↓

Farm business activity areas

Production Production Management Systems (see AREI Chapters 4.8 and 4.9)	Finance Human resources Marketing and contract use

↓

Farm system resource allocation

Crop choice Livestock choice	Other land uses Recreation and woodland

↓

Whole farm system

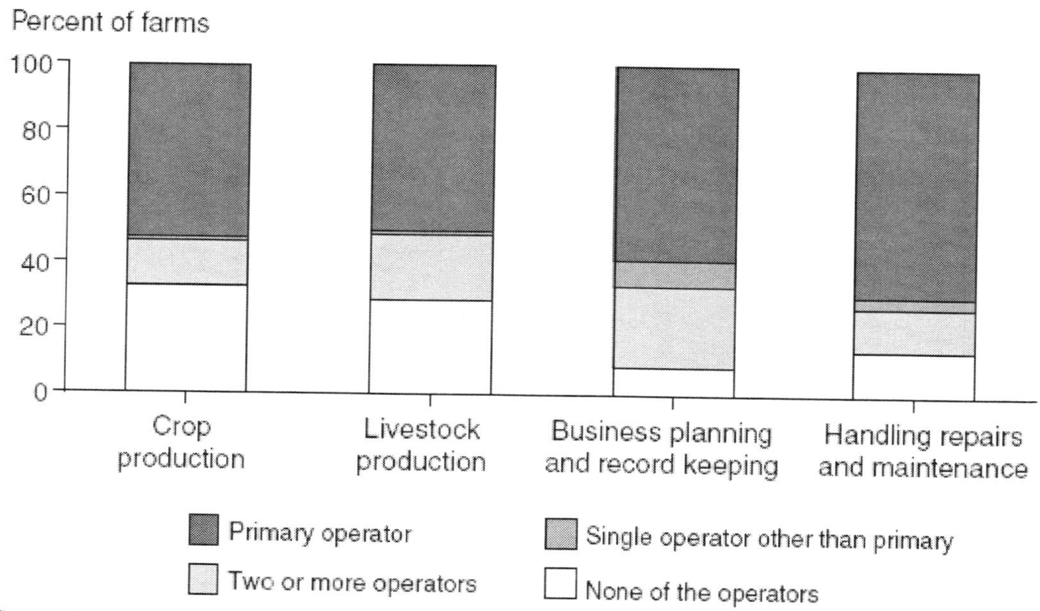

Source: 2003 Agricultural Resource Management Survey.

Figure 38. Participation in farm mangement decisions by operators of farms, 2003.

Table 16. Characteristics of operators and distribution of farms by management unit, 2003

Item	Operator only	Operator with spouse and/or -	Multiple operators	Hired managers	All
Number of farms	773,769	1,076,427	212,773	18,515	2,081,483
Percent of farms	37.2	51.8	10.2	0.9	100.0
Percent of acres operated	16.7	48.7	31.4	3.3	100.0
Percent of production value	14.3	45.2	34.6	5.9	100.0
Average age of operators	57.4	54.9	54.2	56.2	55.8
Primary operators with college (percent)					
Some college	21.4	25.9	28.9	na	24.4
Completed college (BA, BS)	11.1	14.6	15.6	na	13.5
Graduate school	5.8	*6.2	na	na	6.0
Primary operator's major occupation (percent)					
Farm or ranch work	29.1	40.3	48.5	*70.3	37.3
Retired, but still farming	25.3	14.0	*13.4	na	18.1
Work other than farming/ranching	45.6	45.7	*38.2	na	44.6

*Indicates that CV (Coefficient of Variation=(Standard Error/Estimate)*100) is greater than 25 and less than or equal to 50. na indicates value is not available due to no observations, an undefined statistic, or reliability concerns. Rounded percents may not add to 100.

Source: 2003 USDA Agricultural Resource Management Survey

Operator and operator-spouse management teams controlled 59 percent of farms in 2003. When paid or informal advisors are considered, the share rises to 89 percent (Table 16). These farms were, by far, the smallest in terms of acreage and value of production. Management that featured more than two people or outside hired/informal assistance operated larger businesses. Nearly a third of these farms were commercial farms, versus about 3 percent for operator-only units and 7 percent overall. Each management structure that included outside advisors represented a disproportionate share of commercial-size farms.

Operator-only management units had the highest average age, nearly 2 years older than the average for all primary operators. Primary operators in multiple-operator management units were youngest, age 54 on average. Primary operators that used outside assistance had the largest share of college-level attainments. This group was followed by multiple-operator teams and operators in operator-only units.

Farms managed by operators only or by a combination of operators and spouses were more common in the Eastern Uplands, Southern Seaboard, and Mississippi Portal. These regions have a larger share of smaller farms run by operators who work off-farm. Farms run by multiple-operator teams and that used outside assistance were more common in the Heartland, Northern Crescent, Northern Plains, Fruitful Rim, and Basin and Range. Multiple operators were more common on farms that specialized in cash grains and soybeans, high-value crops, and dairy.

Farm Business Management Entails a Host of Choices

Managers of farm businesses make choices about inputs and their use in producing crops, livestock, or other products and services. Production decisions focus on whether to produce crops, livestock, both, or nothing (for example, by placing land in conservation). Financial decisions center on acquiring and maximizing the use of inputs. Do managers have sufficient funds to buy inputs like seed or fertilizer (short-term decisions) or invest in capital items like equipment? If not, is borrowing warranted?

Marketing options range from cash markets to contracts to direct sales (farmers' markets, the Internet, wholesale/retail buyers, or livestock producers.)

Human resource issues include the amount and timing of labor needed to undertake production. In 2003, 45 percent of operators reported their primary occupation as other than farming. An even larger share worked off-farm. Thus, work arrangements vary from self-sufficiency to inclusion of household members, other operators, and a variety of custom hire (person and machine), contract (crew leader), and hired workers. Farmers may even work off-farm and hire someone else to do farm work.

Classifying Farm Business Systems

The result of all the choices across all these business concerns is a highly diverse farm sector. Some farms amount to a single individual supplying all labor to produce one, or maybe even no commodities, with cash sales, and without debt. Other farms produce multiple commodities, market to various outlets, use a variety of labor sources, and take on debt from multiple lenders structured for different periods of maturation.

One way to overcome this complexity empirically is to devise a classification system that jointly considers management choices. To develop the farm business system classification, each of four business areas—production, finance, human resources, and marketing/contract use—was measured as a dichotomous variable (see box, "Classifying Farm Businesses"). For example, farms producing 2 or fewer commodities were assigned a score of zero, while those with 3 or more commodities were given a score of 1. The same scoring convention was used for debt, hired labor, and cash sales versus production or marketing contracts. By equally weighting each of the four business activity areas, a total score ranging from 0 to 1 was calculated to reflect the overall complexity of the operation. For example, a score of 0 indicates a farm having two or fewer commodities, no debt, operator/family labor only, and cash sales.

Classifying Farm Businesses

Farm business activity	Complexity of business organization	
	Lower >>>>>>> > to	> >>>>>>>Higher
Production	2 or fewer commodities	3 or more commodites
Finance	No use of debt	Short- and long-term debt
Human resources	Operator/family labor	Hired labor
Marketing and contract use	Cash sales	Contracts

The scale was used to classify five groups of farms ranging from least (score of 0) to most complex organization (score of 1) based on use of business practices and arrangements. Farms are not distributed equally among the groups. Group 1, for example, accounted for 31 percent of farms in 2003, while Group 5 accounted for 3 percent (Table 17).

CHARACTERISTICS OF FARM BUSINESS SYSTEMS

The 31 percent of farms with the least complex farm business system controlled 11 percent of acres operated and generated less than 3 percent of production value in 2003. Over 85 percent of these farms are rural residences and the rest almost entirely intermediate farms (sales below $250,000 and the operator reports farming as his or her major occupation). Over 98 percent had sales of less than $100,000. These farms specialized in production of field crops other than cash grains or soybeans, beef cattle and general livestock. Operator and operator-spouses managed three-fourths of the least complex farms (Table 18). They had the highest average operator age and the largest share of primary operators over age 65 years (32 percent). Over 80 percent considered their primary occupation to be off-farm and almost 29 percent were retired. This helps explain the 890 hours worked on-farm by the operator, well below the all-farm average of 1,393 hours.

Table 17. Distribution of farm and operator characteristics by complexity of farm business organization, 2003

Item	Least complex>>>>>>Most complex			
	Group 1	Groups 2-4	Group 5	All
Activities/practices employed	0	1-3	4	
Number of farms	648,250	1,366,716	66,517	2,081,483
Percent of farms	31.1	65.6	3.2	100.0
Percent of acres operated	11.0	81.3	7.6	100.0
Percent cash renting land	8.5	31.1	72.4	25.4
Percent of cropland acreage	9.3	76.9	13.7	100.0
Percent hiring labor	0.0	34.7	66	24.9
Average value of production ($)	6,267	89,602	413,358	73,994
Share of production value (%)	2.6	79.4	17.9	100.0
Number of operators	882,603	2,025,299	113,506	3,021,409
Share with one operators (%)	66.7	56.2	41.6	59.0
Share with two operators (%)	31.0	40.4	48.9	37.7
Farm typology				
Rural residence farms (percent)	86.4	61.0	11.1	67.4
Intermediate farms (percent)	13.3	30.6	32.8	25.3
Commercial farms (percent)	na	8.3	56.1	7.3
Operator's major occupation				
Farm or ranch work (percent)	18.3	43.8	87.7	37.3
Average hours primary operator worked on farm	891	1,550	3,068	1,393
Percent with farm financial debt	na	57.1	100.0	40.7
Percent with hired management services	13.5	30.4	53.8	25.9
Percent with informal management team members	5.4	16.9	35.5	13.9

na indicates value is not available due to no observations, an undefined statistic, or reliability concerns.
Rounded percents may not add precisely to 100.
Source: 2003 USDA Agricultural Resource Management Survey.

The most complex farms controlled 8 percent of acres and generated 18 percent of value of production in 2003. Farms in this group were mostly commercial. Over 27 percent had over $500,000 or more in sales (versus 3 percent of all farms). The most complex farms had a much larger share of management teams that included multiple persons. These farms were more common in the Northern Crescent, Heartland, Northern Plains, and Eastern Uplands. They tend to specialize in dairy, poultry, hogs, cash grains, and soybeans. Nearly two-thirds reported hiring other individuals and three-fourths had custom hire assistance. The primary operators in these management units were younger, averaging 49 years, nearly 7 years less than the all-farm average. On average, primary operators reported working over 3,000 hours on their farms in 2003, with spouses and other operator labor adding more than 1,100 hours to the total.

Table 18. Distribution of farms by farm management team and complexity of farm business organization, 2003

Item	Least complex>>>>>>Most complex			
	Group 1	**Groups 2-4**	**Group 5**	**All**
Activities/practices employed	0	1-3	4	
Number of farms	648,250	1,366,716	66,517	2,081,483
		Percent of farms		
Composititon of farm management teams				
Operator only	49.1	32.6	16.0	37.2
Operators with advisors	10.5	17.7	20.6	15.6
Operator and spouse	27.1	21.9	14.7	23.3
Operator and spouses with advisors	6.4	15.2	27.6	12.9
Multiple operators	*5.9	5.1	6.0	5.4
Multiple operators with advisors	na	6.3	14.0	4.8
Hired managers	na	#1.0	na	*0.9

Coefficient of Variation = (Standard Error/Estimate) x 100. * indicates that CV is greater than 25 and less than or equal to 50.

# indicates that CV is greater than 50 and less than or equal to 75.

na indicates value is not available due to no observations, an undefined statistic, or reliability concerns.

Rounded percents may not add precisely to 100.

Source: 2003 USDA Agricultural Resource Management Survey.

SUMMARY

Farm managers not only have to be highly skilled at the technical aspects of farm production, but they also have to handle primary and support activities for their farms that range from input procurement to technology, finance, accounting, and human resource management (Gray et al.). Changes in crop and livestock production, including use of contract arrangements and technologically modified seed stock, mean that managers may need to interact more with both suppliers and customers. With the rising cost of inputs, particularly capital items such as machinery and equipment, managers also have to control a range of financial arrangements that transcend farm mortgages. Even land rents have become more complex, with some arrangements incorporating changes in prices and yields.

As a farm grows, expertise to handle these tasks either has to be available within the existing owner-operator-management arrangement or be acquired by adding to the management team. Survey results suggest that the size and composition of management teams align with the complexity of farm businesses. The least complex farms were most often managed by a single operator or by a combination of operator and spouse. Conversely, the most complex businesses typically involved operators, spouses, other partners, and outside advisors. Many Federal and State programs provide income and technical/other assistance to farmers, specifically a farm's decisionmaker. In today's farm sector, that person is not automatically the farm operator alone, especially on farms with the most cropland and production.

REFERENCES

[1] Gray, Allan, Michael Boehlje, Craig Dobbins, and Cole Ehmke. *The Internal Analysis of Your Farm Business.* Purdue Extension, Purdue University, West Lafayette, Indiana, EC-721.

[2] U.S. Department of Agriculture, National Agricultural Statistics Service (1999). *Agricultural Economics and Land Ownership Survey.* Volume 3, Part IV Special Studies of 1997 Census of Agriculture, Table 57.

[3] U.S. Department of Agriculture, National Agricultural Statistics Service (2004). *2002 Census of Agriculture.* Volume 1, Geographic Area Series, Part 51, Table 61.

In: Agricultural Resources and Environmental Indicators ISBN 978-1-60021-467-7
Editor: Keith Wiebe and Noel Gollehon, pp. 117-128

Chapter 13

SOIL MANAGEMENT AND CONSERVATION

Carmen Sandretto and James Payne

Soil quality is critical for plant growth, and therefore important to agriculture and rural ecosystems. Management practices that are appropriate for local soil characteristics and climate can enhance soil quality. These beneficial practices include crop rotations, crop residue management (including cover crops and conservation tillage), and various field/landscape structures and buffers. Crop residue management is generally a cost-effective method of erosion control. It usually maintains or increases crop yields, but requires fewer resources than intensive structural measures and can be implemented in a timely manner to meet conservation needs.

INTRODUCTION

Crop production and its environmental effects depend on the quality of soil. Soil provides the physical, chemical, and biological processes required to sustain most terrestrial plant and animal life. Soil regulates water flow from rainfall, snowmelt, and irrigation between infiltration, root-zone storage, deep percolation, and runoff (National Research Council, 1993). Soil acts as a buffer between production activities and the environment by facilitating the cycling and decomposition of organic wastes and nutrients (carbon, nitrogen, phosphorus, and others), as well as the degradation of nitrates, pesticides, and other toxic substances that are potential pollutants in water or air (Kemper et al., 1997). Soil quality determines how well soil performs its functions.

Soil has both inherent and dynamic qualities. Inherent qualities are those factors, such as texture, that affect a soil's natural ability to function, but do not change easily. Dynamic qualities depend on how a soil is managed. Soils respond differently to management, depending on the inherent properties of the soil and the surrounding landscape. Traditional measures of soil quality include land capability and suitability, productivity, erodibility, and vulnerability to leach pesticides and nitrates (Karlen et al., 1997). A comprehensive soil quality measure would combine these physical attributes with broader societal concerns, such as potential surface-water pollution from field runoff, protecting long-term soil productivity, and the health of agricultural/rural ecosystems.

Soil quality can be maintained or enhanced through the use of appropriate crop production technologies and related resource management systems that involve the composition, structure, and function of entire ecosystems. Beneficial farm-level soil management practices are designed to maintain the quality and long-term productivity of the soil and to mitigate environmental damage from crop production. These practices include crop rotations, crop residue management (including cover crops) and conservation tillage, and field/landscape scale engineering structures and buffers like grass waterways, terraces, contour-farming, strip-cropping, underground drainage outlets, and surface diversion/drainage channels. Also beneficial to soil quality are certain nutrient (see Part 4), pest (see Part 4), and irrigation practices (see Part 4).

The appropriateness of soil management technologies depends on topographic and agro-climatic conditions; site-specific technical, economic, and financial feasibility; farmer attitudes, perceptions, and resources; and society's attitudes toward the range of offsite effects associated with agricultural production (USDA, 1997). Soil management practices can enhance soil quality by:

- Increasing ground cover and organic matter,
- Tilling sparingly to reduce organic matter degradation and compaction,
- Managing fertilizer and pesticide use to minimize their impact on nontarget organisms and water/air quality, and
- Increasing the diversity of plants, wildlife, and other organisms to help control pest populations.

Crop Rotation Systems

Crop rotation (see box, "Cropping Pattern Definitions") can help conserve soil, maintain its fertility, and control pests, diseases, harmful insects, and weeds. Rotating high-residue and/or closely grown crops with row crops can reduce soil losses on erodible soils. Closely grown field grain crops—such as wheat, barley, and oats, as well as hay and forage crops—provide vegetative cover to reduce soil erosion and water runoff while adding organic matter. In addition, these crops help to control broadleaf weeds and may help control weed infestation in subsequent crops. Crop rotation also helps to break disease and insect cycles. Leguminous crops can increase nitrogen levels in the soil, and cover crops planted in the fall help reduce erosion from winter and spring storms, hold nutrients that might otherwise be lost, enhance the soil's biological processes, and lengthen periods of active plant growth (to increase nutrient cycling, disease suppression, soil aggregation, and carbon sequestration).

Crop Rotation System Use For Major Crops

With the exception of cotton, rotational cropping in some form dominates major crop production in the United States. The most common rotation system for both corn and soybeans is a corn-soybean rotation. This combination reduces erosion (compared with continuous-corn or continuous-soybeans); helps control disease, insects, and weeds; and enables soybeans to fix nitrogen for use by the subsequent corn crop. Approximately 75 percent of corn acres and 80 percent of soybean acres in the 10 major producing States used this rotation system in the most recent surveyed year (2001 for corn and 2002 for soybeans) (Figures 39 and 40).

Cropping Pattern Definitions

The following definitions were applied to 3-year crop sequence data reported in the Agricultural Resource Management Survey to identify a cropping pattern for each sample field. The data were limited to the current year's crop plus the crops planted the previous 2 years on the sample field, with the exception of winter wheat in 1996. For this crop, only 2 years were used to determine the rotation due to data limitations.

Monoculture or continuous same crop: crop sequence where the same crop is planted for 3 consecutive years. Small grains (wheat, oats, barley, flax, rye, etc.) or other close-grown crops may be planted in the fall as a cover crop.

Continuous row crops: crop sequence, excluding continuous same crop, where only row crops (corn, sorghum, soybeans, cotton, peanuts, vegetables, etc.) are planted for 3 consecutive years. Small grains or close-grown crops may be planted in the fall as a cover crop.

Continuous small grain crops: crop sequence, excluding continuous same crop, where only small grain crops (wheat, barley, oats, rye, etc.) are planted for 3 consecutive years

Row crop/small grain rotation: crop sequence where some combination of row crops and small grains are planted over the 3-year period.

Rotation with meadow crops: crop sequence that includes hay, pasture, or other use in 1 or more previous years. The rotation excludes any of the above rotations and any area that was idle or fallow in one of the previous years.

Idle or fallow in rotation: crop sequence that includes idle, diverted, or fallowed land in 1 or more of the previous years.

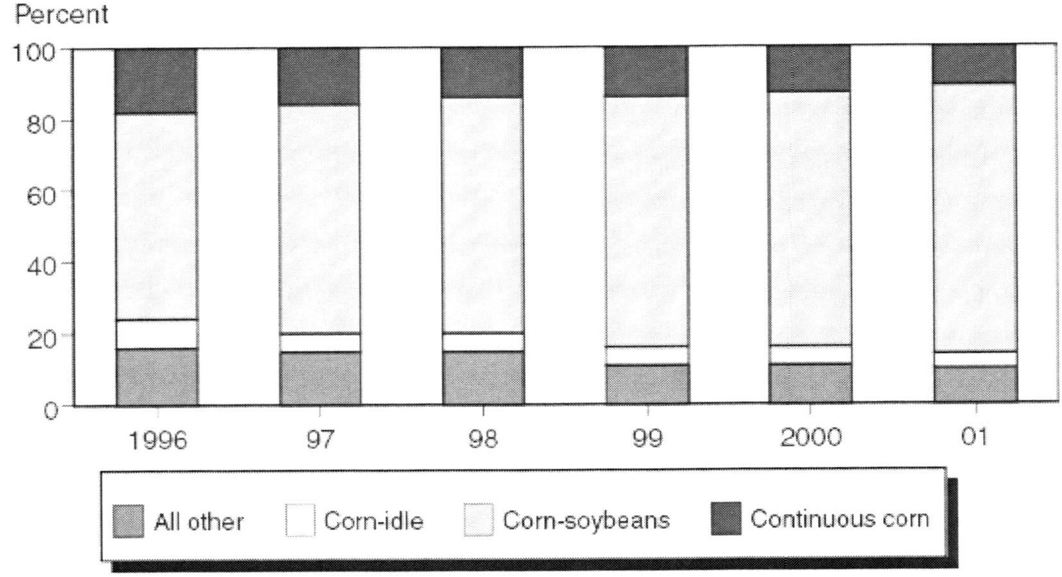

Source: USDA, ERS, Agricultural Resource Management Surveys.

Figure 39. Cropping patterns on corn for 10 major production States, 1996-2001.

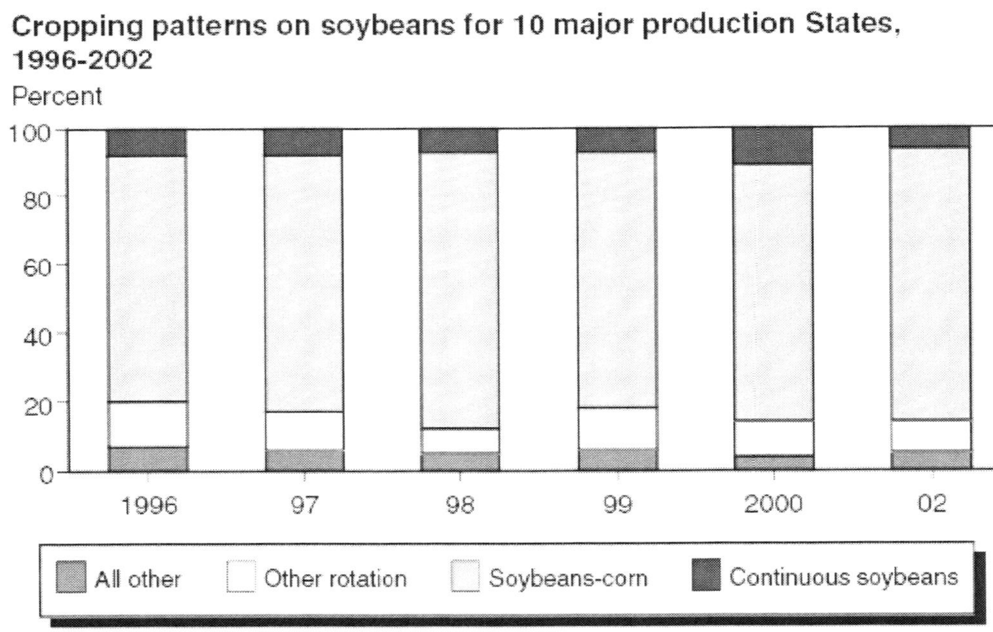

Source: USDA, ERS, Agricultural Resource Management Surveys.

Figure 40. Cropping patterns on soybeans for 10 major production States, 1996-2002.

Winter wheat in a continuous cropping system (no rotation) reached a high of 47 percent of acreage planted in 2000 (most recent surveyed year) (Figure 41). Winter wheat in rotation with a row crop or small grain (including double cropping) has trended upward in recent years, while rotation with fallow/idle has declined. Cotton is grown primarily in a continuous cropping system, with 73 percent of acreage in the five major States using this system in 2003. The most common cotton rotation was cotton-row crop at around 20 percent (Figure 42).

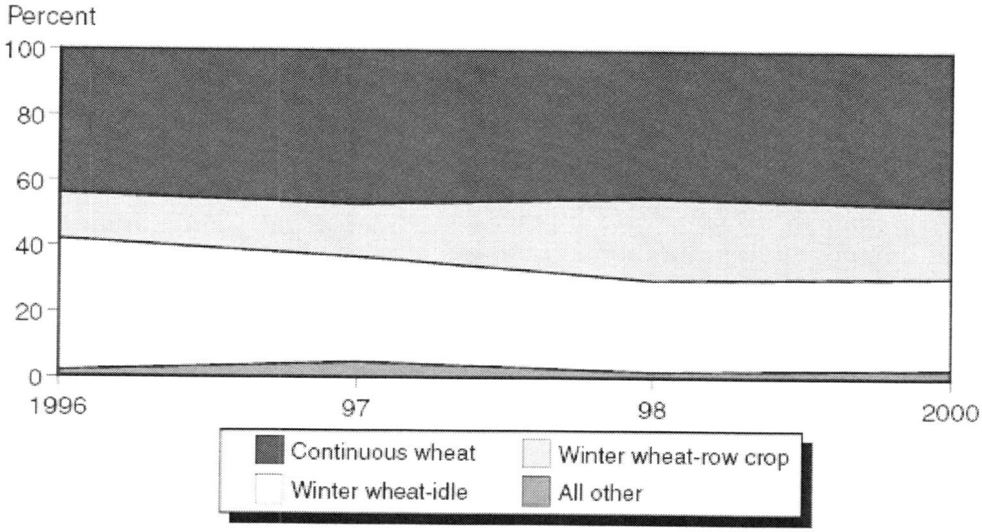

Source: USDA, ERS, Agricultural Resource Management Surveys.

Figure 41. Cropping patterns on winter wheat for 10 major production States, 1996-2000.

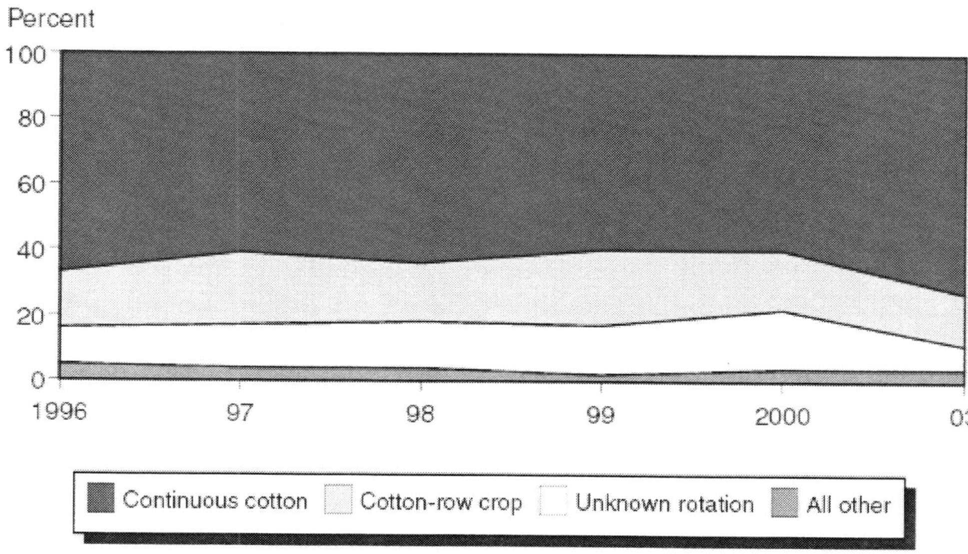

Source: USDA, ERS, Agricultural Resource Management Surveys.

Figure 42. Cropping patterns on cotton for 5 major production States, 1996-2003.

Economic Factors Affecting Farmers' Choices

A farmer chooses a cropping pattern based mostly on the relative rate of return resulting from differences in yields, costs and returns, and government policy. Crop rotations usually result in yields higher than those achieved with continuous cropping under similar conditions. Rotations that add organic matter can improve soil tilth and water-holding capacity, and thus increase crop yields. Grain yields following legumes are often 10 to 20 percent higher than continuous grain, regardless of the amount of fertilizer applied (Heichel, 1987; Power, 1987). Corn following wheat produces a greater yield than continuous-corn with the same amount of fertilizer, even though wheat is not a legume and cannot fix atmospheric nitrogen (Power, 1987). Rotations with legumes can increase available soil nitrogen and reduce the need for commercial nitrogen fertilizers. Legumes in a rotation are most effective in humid and sub-humid climates where they do not decrease subsoil moisture for subsequent crops.

Crop rotations—by alternating a susceptible crop with a nonhost crop—can help to control a variety of pests by disrupting their life cycles. Soil microbiology and beneficial insects thrive under crop rotations, and this helps control disease and other pests, particularly those that attack plant roots. For example, rotating corn with soybeans can reduce the need for insecticide treatment when the field is in corn by reducing the number of corn root-worm larvae in the soil (although the effectiveness of this practice may be decreasing in some areas).

The diversification inherent in rotations can be an economic buffer against fluctuating prices of crops or production inputs and against the vagaries of weather, disease, and pest infestations.

Policies and Programs Affecting Cropping Patterns

Federal policies influence farmers'choices of crops and management practices. Past commodity programs that restricted base acreage to program crops encouraged monoculture or continuous planting of the same crop. Starting with the 1990 Food, Agriculture, Conservation and Trade Act, farmers were given the option to diversify (without incurring a penalty) their program crop base acres. Farmers began to grow other crops and/or use rotations in response to changes in prices and loan deficiency payments.

Under the 1985 Food Security Act and subsequent farm legislation, highly erodible land (HEL) used for crops required implementation of a conservation plan in order to be eligible for USDA farm program benefits (see Chapter 23, Compliance Provisions for Soil and Wetland Conservation). Rotating row crops with less erosive crops such as small grains and hay/pasture is a key part of some conservation plans for HEL, usually in combination with cover crops, crop residue management, and conservation tillage.

Crop Residue Management

Crop residue management (CRM) maintains additional crop residue on the soil surface through fewer and/or less intensive tillage operations. CRM is generally cost effective in protecting soil and water resources and can lead to higher returns by reducing fuel, machinery, and labor costs while maintaining or increasing crop yields, but requires fewer resources than intensive structural measures and can be implemented in a timely manner to meet conservation needs (USDA, 1997). CRM systems include reduced tillage, conservation tillage (no-till, ridge-till, and mulch-till), and the use of cover crops and other conservation practices that leave sufficient residue to protect the soil surface from the erosive effects of wind and water (see box, "Crop Residue Management and Tillage System Definitions").

Why Manage Residue?

Historically, crop residues were removed from farm fields for livestock bedding, feed, or sale. Residues that remained on the field were burned off to control pests, plowed under, or tilled into the soil. Culturally, some farmers would take pride in having their fields "clean" of residue and intensively tilled to obtain a smooth surface in preparation for planting. More recently, farmers have adopted CRM practices—with government encouragement—because of new knowledge about residue's benefits and improved planters, crop protection technologies, and the like (USDA, 1997).

CRM can benefit society through enhanced environmental quality and farmers through higher overall economic returns. However, adoption of CRM may not lead to clear environmental benefits in all regions and may not be profitable on all farms. Public and private interests support cooperative efforts to address the barriers to realizing greater benefits from CRM practices. For example, recent advances in planting equipment permit Seeding new crops through heavier surface residue into untilled soil and even directly into killed sod (USDA, 1997).

Trends In Crop Residue Management Use

According to the Conservation Technology Information Center's National Crop Residue Management Survey, U.S. farmers practiced CRM on about 172 million acres in 2004, or 62 percent of planted acreage, up from 144 million acres in 1990. Conservation tillage accounted for 41 percent of U.S. planted crop acreage in 2004, compared with 26 percent in 1990. Most of the growth in conservation tillage since 1990 has come from expanded adoption of no-till (Figure 43), which can leave 70 percent or more of the soil surface covered with crop residue. U.S. crop area planted with no-till more than tripled from 17 million acres (6 percent) to 62 million acres (22 percent) between 1990 and 2004 (CTIC, 2005).

Economic Incentives For CRM Adoption

Yield response with soil-conserving tillage systems varies with location, soil characteristics, climate, cropping patterns, and level of management skills (Sandretto, 2001). In general, long-term field trials on well-drained to moderately well-drained soils or on sloping land show slightly higher no-till yields, particularly with crop rotations, compared with intensive tillage (CTIC, 1996). Benefits from improved moisture retention in the root zone usually increase crop yields, especially under dry conditions. In some areas, these benefits permit a change in the cropping pattern to reduce the frequency of moisture-conserving fallow periods (USDA, 1997). Other benefits derive from more timely preparation for double cropping, with better yields as one result.

Crop Residue Management and Tillage System Definitions

Unmanaged	Crop Residue Management (CRM)			
Intensive- or conventional-till	Reduced-till	Conservation tillage		
		Mulch-till	Ridge-till	No-till
Moldboard plow or other intensive tillage used	No use of moldboard plow and intensity of tillage reduced	Full-width tillage, but further decrease in tillage intensity	Only the tops of ridges are tilled	No tillage performed since harvest of previous crop
<15% residue cover remaining	15-30% residue cover remaining	30% or greater residue cover remaining on soil surface after planting		

Crop Residue Management (CRM)—A year-round system that usually involves a reduction in the number of passes over the field with tillage implements and/or in the intensity of tillage operations, including the elimination of plowing (inversion of the surface layer of soil). CRM begins with the selection of crops that produce sufficient quantities of residue to reduce wind and water erosion and may include the use of cover crops after low-residue-producing crops. CRM is an umbrella term encompassing several tillage systems including conservation tillage (no-till, ridge-till, and mulch-till), and reduced-till. (Note: reduced-till is not considered a part of conservation tillage.)

Conservation tillage—Any tillage and planting system that maintains at least 30 percent of the soil surface covered by residue after planting to reduce soil erosion. Two key factors influencing crop residue are: (1) the type of crop, which establishes the initial residue amount and its fragility, and (2) the type of tillage operations prior to and including planting. No-till, ridge-till and mulch-till are thee common types of conservation tillage systems.

No-till—Residue from the previous crop is undisturbed from harvest to planting except for nutrient injection or narrow strips. Weed control is primarily accomplished with crop protection products.

Ridge-till—Residue from the previous crop is undisturbed from harvest to planting except for nutrient injection. Planting is completed in a seedbed prepared on 4- to 6-inch high ridges that are formed and rebuilt during row cultivation for weed control. Residue is left on the surface between ridges.

Mulch-till—A full-width tillage system usually involving one to three tillage passes over the field performed prior to and/or during planting, that leaves, after planting, at least 30 percent of the soil surface covered with residue.

Reduced-till (15-30% residue)—Full-width tillage usually involving one or more tillage passes over the field performed prior to and/or during planting, that leaves 15-30 percent residue cover after planting.

Conventional-till or intensive-till (less than 15% residue)—Full-width tillage that is performed prior to and/or during planting, that generally involves plowing with a moldboard plow and/or other intensive tillage equipment. Less than 15 percent residue cover remains on the soil surface after planting.

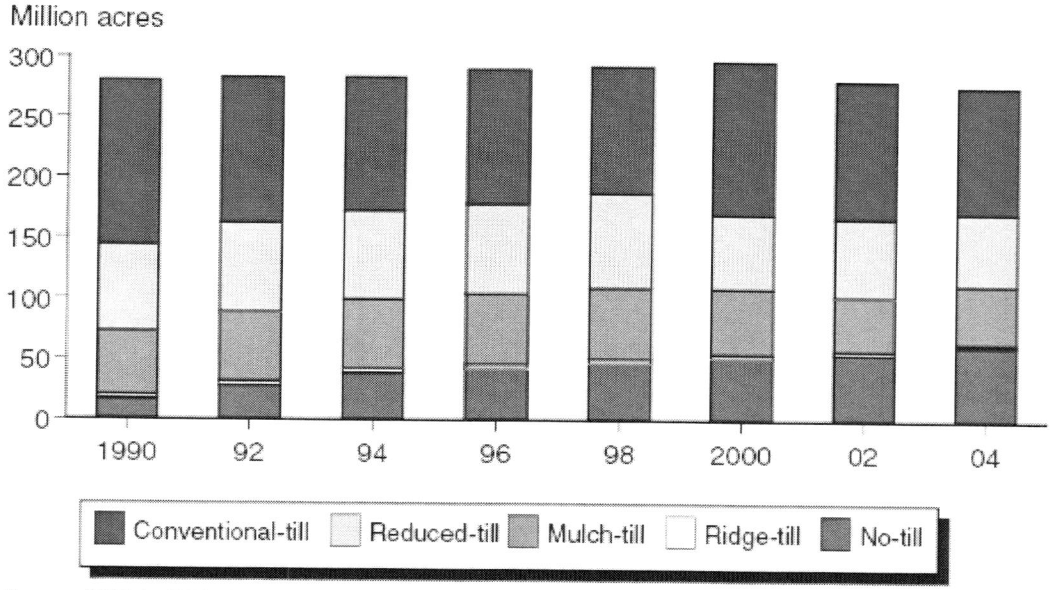

Source: USDA, ERS, based on National Crop Residue Management Survey data from the Conservation Technology Information Center (CTIC).

Figure 43. Tillage types, 1990-2004.

Crop yields can be significantly reduced by pest populations, which frequently change under different tillage systems and are also affected by cropping pattern. Maintaining or increasing yields when changing tillage systems requires skillful use of the various means of pest control, including crop variety selection, proper application of crop protection products, row cultivation, cover crops, crop rotation, scouting, and other integrated pest management practices (see Chapter 14, "Pest Management Practices"). Use of crop protection products on major crops differs among tillage systems, but the effects related to tillage systems are difficult to distinguish from differences in pest populations due to other factors, including use of other pest control practices (USDA, 1997).

Choice of tillage system affects machinery, chemical, fuel, and labor costs. Decreasing the intensity of tillage and/or reducing the number of tillage operations (fewer trips over the field) reduces labor requirements per acre, extends equipment life, increases the area covered, and reduces fuel and maintenance costs. These cost savings may be offset by increased crop protection costs and the fertilizers required to attain optimal yields (Sandretto, 2001). Conservation tillage may increase net returns on the entire farming operation even if returns for a particular crop do not increase. For example, a tillage system that requires substantially less labor per acre and reduces returns per acre only slightly may free up labor to serve more acres or generate more income elsewhere (Sandretto and Bull, 1996).

Potential Environmental Benefits of Crop Residue Management

Soil quality can benefit from minimum tillage and maximum residue, and this combination contributes to improved ecosystem health in several ways.

Tillage systems that leave substantial amounts of crop residue evenly distributed over the soil surface **reduce soil erosion**, from reduced wind erosion and reduced kinetic impact of rainfall, surface sediment transport and water runoff; with increased water infiltration and moisture retention (Edwards, 1995). Several field studies conducted on small watersheds under natural rainfall on highly erodible land have shown that erosion rates with the moldboard plow can be reduced by 70 percent or more with conservation tillage (USDA, 1997).

Surface residues help intercept nutrients and chemicals and hold them in place until they are used by the crop or degrade into harmless components, which provide **cleaner surface runoff** (USDA, 1997). Increased organic matter in the top layer of soil results in cleaner runoff, and thus benefits water quality by reducing the flow of contaminants such as sediment and adsorbed/dissolved chemicals into lakes and streams (USDA, 1997; CTIC, 1996). Studies under field conditions indicate that while the quantity of water runoff from no-till fields was variable depending on the frequency and intensity of rainfall, clean-tilled soil surfaces produce substantially more runoff (Edwards, 1995). Average herbicide runoff losses from treated fields under no-till and mulch-till systems for all products and all years were about 30 percent of the runoff levels from moldboard-plowed fields (Fawcett et al., 1994).

Crop residues on the soil surface, by creating tiny dams, enhance infiltration, reduce surface-crust formation, and slow water runoff, which **increases water infiltration and soil moisture** (Edwards, 1995). The channels (macropores) created by earthworms and old plant roots, when left intact with no-till, improve infiltration to help reduce or eliminate field runoff and provide water quality benefits. Combined with reduced water evaporation from the top few inches of soil and with improved soil characteristics, the higher level of soil moisture can contribute to higher crop yields in many cropping and climatic situations (CTIC, 1996).

Less intensive tillage reduces breakdown of crop residue and loss of soil organic matter **improving long-term soil quality**. Carbon sequestration may increase to build soil organic matter, enhance biological (including earthworm) activity, and maintain long-term productivity. Conservation tillage, particularly continuous no-till, improves soil structure by increasing soil particle aggregation (small soil clumps), aiding water movement through the soil so plants expend less energy to establish roots. No-till also reduces soil compaction through fewer trips over the field and reduced equipment weight and horsepower requirements (CTIC, 1996).

These potential environmental benefits suggest a **public role** in encouraging adoption of crop residue management practices. Conservation compliance provisions of the 1985 Food Security Act and subsequent farm legislation have given farmers additional incentives to adopt CRM to control erosion (and thereby improve water quality), particularly on highly erodible crop-land (HEL) (see Chapter 23, "Compliance Provisions for Soil and Wetland Conservation"). Expanded use of CRM practices on non-HEL indicates that producers are motivated by the potential to reduce costs, improve efficiency, and/or increase soil productivity.

Conservation Buffers and Structures

Soil and water conservation structures and buffer zones can significantly reduce erosion and sediment transport caused by rainfall and water runoff. These structures allow for surface water to be captured onsite or slowed and diverted from the field via erosion-resistant waterways, channels, or outlets. While management practices, such as crop rotation, crop residue management (including cover crops), and conservation tillage practices help to control erosion, they may not sufficiently control runoff water after heavy rainfall events. Soil- and water-conserving structures, therefore, are important in farm soil management systems. Engineering structures and buffer zones for soil and water conservation vary significantly across crop production regions to reflect the wide variation in soil, climate, and cropping patterns.

A variety of USDA programs since the 1930s have provided cost-sharing and technical assistance for conservation buffers, structures, and practices (see Chapter 24, "Working-Land Payment Programs"). While recent program efforts have been directed toward management practices, including vegetative cover establishment and crop residue management, some cropland continues to be served by installation of terraces and other structural measures to better control sediment and water runoff.

REFERENCES

[1] Conservation Technology Information Center (2005). *National Crop Residue Management Survey, Conservation Tillage Data*, West Lafayette, IN.

[2] Conservation Technology Information Center (1996). *A Checklist for U.S. Farmers*, West Lafayette, IN.

[3] Edwards, William M. (1995). "Effects of Tillage and Residue Management on Water for Crops," in *Crop Residue Management to Reduce Erosion and Improve Soil Quality: Appalachia and Northeast Region*, R.L. Blevins and W.C. Moldenhauer (eds.), U.S. Dept. Agr., Agr. Res. Serv., Conservation Research Report No. 41.

[4] Fawcett, Richard S., Dennis P. Tierney, Brian R. Christensen (1994). "Impact of Conservation Tillage on Reducing Runoff of Pesticides into Surface Waters," *Journal of Soil and Water Conservation*, Vol. 49, No. 2.

[5] Heichel, G.H (1987). "Legumes as a source of nitrogen in conservation tillage systems," *The Role of Legumes in Conservation Tillage Systems*, J.F. Power (ed.), Ankeny, IA: Soil Conservation Society of America.

[6] Karlen, D.L., M.J. Mausbach, J.W. Doran, R.G. Cline, R.F. Harris, and G.E. Schuman (1997). "Soil Quality: A Concept, Definition, and Framework for Evaluation," *Soil Science Society of America Journal*, 61(1).

[7] Kemper, W.D., E.E. Alberts, C.D. Foy, R.B. Clark, J.C. Ritchie, and R.W. Zobel (1997). "*Aerenchyma*, acid tolerance, and associative nitrogen fixation enhance carbon sequestration in soil," in R. Lal, J.M. Kimble, R.F. Follett, and B.A. Stewart (eds.) *Management of Carbon Sequestration in Soil.* CRC Press, Boca Raton, FL.

[8] Power, J.F (1987). "Legumes: Their Potential Role in Agricultural Production." *American Journal of Alternative Agriculture* 2(2): 69-73.

[9] National Research Council (NRC) (1993). *Soil and Water Quality: An Agenda for Agriculture*, National Academy Press, Washington, DC.

[10] Sandretto, C., and L. Bull (1996). "Conservation Tillage Gaining Ground," *Agricultural Outlook: Resources and Environment, Special Article*, U.S. Dept. Agr., Econ. Res. Serv., AO-232, Aug.

[11] Sandretto, C.L. (2001). "Conservation Tillage Firmly Planted in U.S. Agriculture," *Agricultural Outlook*, U.S. Dept. Agr., Econ. Res. Serv., AGO-279, March.

[12] USDA [U.S. Department of Agriculture], Economic Research Service (1997). Chapter 4 "Crop Residue Management," *Agricultural Resources and Environmental Indicators, 1996-97*, AH-712. July.

In: Agricultural Resources and Environmental Indicators ISBN 978-1-60021-467-7
Editor: Keith Wiebe and Noel Gollehon, pp. 129-138 © 2007 Nova Science Publishers, Inc.

Chapter 14

PEST MANAGEMENT PRACTICES

Craig Osteen and Michael Livingston

Crop producers use pesticides and other practices to manage pests. The quantity of pesticides used on crops was less in 2002 than in 1997. The development and marketing of new pesticides and the adoption of genetically modified seed have resulted in changes to insecticide and herbicide compounds used, while new pest problems can increase pesticide use.

INTRODUCTION

Crop producers use pesticides and other practices to manage insects, diseases, and weeds and to prevent crop yield or quality losses. Factors that influence pest management decisions include the extent of pest problems, cost and effectiveness of available practices, regulations on what pesticides can be used and how, and the prices of commodities and inputs. The recent entry of Asian soybean rust into the United States could increase fungicide use.

PESTICIDE USE

Pesticide use can be measured by expenditures, quantity, and area of use. The measure for which estimates have been available for the longest time is million pounds of active ingredient (a.i.). However, this measure does not capture changes in the use of pesticide compounds applied at different rates (where the area treated is unchanged). Nor is total quantity a good measure of total pesticide toxicity, which varies by pesticide compound, or of risk, which can be mitigated by application practices.

One measure of pesticide area is acre-treatments, which is the product of acreage treated and treatments per acre. We use this measure when discussing market shares of insecticides and herbicides, because some pesticides are applied at low rates per acre and account for small portions of total quantity applied, but large portions of total treatments.

Agricultural pesticide expenditures reached an estimated all-time high of $9 billion in 1997-98 and totaled $8.3–8.5 billion in 2002-2004. Herbicides accounted for two-thirds of those expenditures, while insecticides accounted for about one-fifth in 2000 and 2001 (Kiely et al., 2004). Crop pesticide use peaked at an estimated 579 million pounds a.i. in 1997; in 2004, it was 495 million pounds a.i. (Figure 44, Table 19).

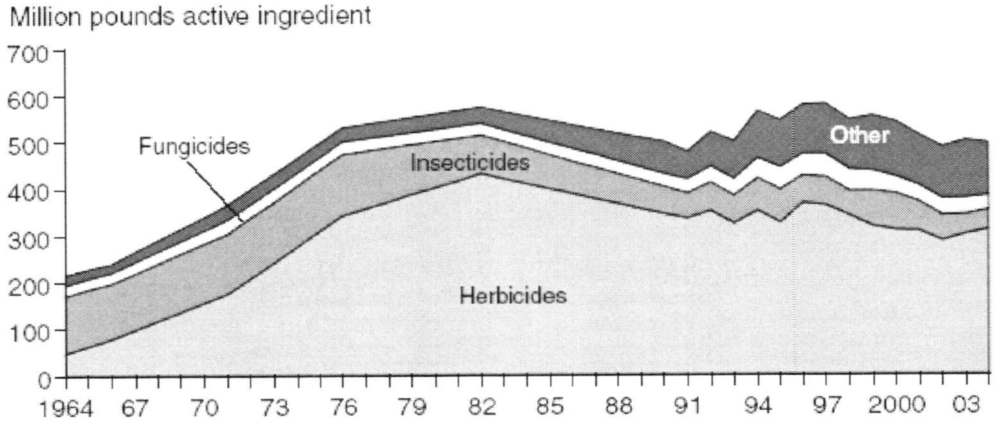

[1]Linear interpolation of use estimates between survey years from 1964 to 1990.
Source: Padgitt et al., 2000, U.S. Census Bureau and unpublished ERS data.

Figure 44. Pesticide use on major crops, 1964-2004[1].

Table 19. Quantity of pesticides applied, total and to selected crops, 1964-2004

Type of pesticide and commodity	1964	1971	1982	1991	1997	2004
	Quantity of pesticides applied (million pounds active ingredient)					
Total	215.0	364.4	572.4	477.5	579.3	494.5
Herbicides	48.2	175.7	430.3	335.2	362.6	311.0
Insecticides	123.3	127.7	82.7	52.8	60.2	40.7
Fungicides	22.2	29.3	25.2	29.4	48.5	29.8
Other	21.4	31.7	34.2	60.1	108.0	112.9
Corn	41.2	127.0	273.7	233.2	227.3	174.6
Cotton	95.3	111.9	49.5	50.3	68.4	56.7
Wheat	10.1	13.6	23.5	13.8	25.5	22.3
Soybeans	9.2	42.2	147.4	70.4	83.5	87.8
Potatoes	6.1	15.5	24.6	35.6	59.4	62.1
Other vegetables	20.8	20.7	21.7	40.3	73.3	65.1
Citrus fruit	8.1	14.1	16.5	13.7	15.0	7.2
Apples	19.9	12.7	10.0	9.1	10.6	8.5
Other deciduous fruit	4.4	6.6	5.5	11.1	16.4	10.3

Sources: Padgitt et al., 2000; U.S. Census Bureau; and unpublished ERS data.

In recent decades, the development of new pesticides, increased use of practices such as genetically modified seed, and the regulatory process encouraged shifts in pesticide compounds used. Many, but not all, compounds increasing in total use are applied at lower rates per acre than those declining in total use, resulting in lower pesticide quantity. Rather than discuss hundreds of pesticide compounds, we show use of major insecticide and herbicide families measured by shares of total quantity applied and acre-treatments on five major crops: corn, cotton, potatoes, soybeans, and wheat.

Among insecticides, the organophosphate share of acre-treatments and quantity applied was greater in 2000 than 1996, while pyrethroid and carba-mate shares were less (Table 20). Corn and cotton are the major insecticide markets among the five crops. The higher organophosphate share was largely due to higher malathion use on cotton for boll weevil eradication, which has since declined. During 1996-2000, organophosphate and pyrethroid use on corn varied year to year, with no obvious trend. (See "Crop Production Practices" in the ARMS Data Tool on the ERS website.)

Table 20. Shares of insecticide use by family, 1964-2000[1]

Insecticide family	1964	1971	1982	1991	1996	2000
				Percent		
Quantity						
Carbamates[2]	7	10	15	11	12	7
Organochlorines[3]	73	51	9	2	3	*
Organophosphates[4]	20	39	71	80	80	86
Pyrethroids[5]	0	0	4	3	4	2
Others	0	0	*	5	2	5
Acre-treatments[6]						
Carbamates	NA	NA	14	11	10	8
Organochlorines	NA	NA	5	2	1	*
Organophosphates	NA	NA	60	57	54	60
Pyrethroids	NA	NA	21	27	29	20
Others	NA	NA	*	3	6	12

NA = Not available.

* = Less than 1 percent.

[1]Estimated for corn, cotton, potatoes, soybeans, and wheat; excludes oils, sulfur, and other inorganics. Since potatoes were not surveyed in 2000, the 2000 estimate includes potato use in 1999.

[2]Examples include aldicarb, carbaryl, carbofuran, formetanate, methomyl, and oxamyl.

[3]Examples include dicofol, endosulfan, methoxychlor, and many materials no longer registered: aldrin, chlordane, deldrin, DDT, and toxaphene. [4]Examples include azinphos-methyl, chlorpyrifos, fonodos, malathion, methyl parathion, mevinphos, parathion, phorate, and terbufos.

[5]Examples include permethrin, cypermethrin, tralomethrin, deltamethrin, cyhalothrin, cyfluthrin, and esfenvalerate.

[6]Sum of acreage treated with a pesticide multiplied by average number of applications per acre.

Source: Eichers et al., 1968; Andrelenas, 1974; unpublished ERS data.

According to NASS, malathion was used on 11 percent of cotton acres (with 6 treatments per acre) in 1997 and 1998, 40 percent (7 treatments per acre) in 1999, but only 11 percent (5 treatments per acre) in 2003. The U.S. Environmental Protection Agency (USEPA), under a regulatory review of organophosphates (discussed below), determined that malathion use for boll weevil eradication was not a significant dietary and drinking water health risk (USEPA, 2000). The adoption of cotton seed genetically modified to produce the *Bacillus thuringiensis* toxin may reduce organophosphate, pyrethroid, and carbamate insecticide use for lepidopteran insects, such as bollworms and tobacco budworms (see Chapter 9, "Biotechnology and Agriculture").

Table 21. Shares of herbicide use by family, 1964-2000[1]

Herbicide family	1964	1971	1982	1991	1996	2000
			Percent			
Quantity						
Amides[2]	0	24	31	35	33	28
Anilines[3]	2	8	11	12	13	9
Carbamates[4]	10	5	17	9	4	*
Phenoxys[5]	43	12	4	4	7	4
Triazines[6]	23	32	26	29	27	22
Phosphinic acids[7]	0	0	1	2	5	23
Sulfonyl ureas[8]	0	0	*	*	*	*
Other new families[9]	0	0	3	3	8	8
Others	22	16	6	6	4	4
Acre-treatments[10]						
Amides	NA	NA	20	16	12	11
Anilines	NA	NA	15	13	10	6
Carbamates	NA	NA	6	2	1	*
Phenoxys	NA	NA	13	10	12	7
Triazines	NA	NA	26	24	18	14
Phosphinic acids	NA	NA	1	2	6	20
Sulfonyl ureas	NA	NA	*	9	13	13
Other new families[9]	NA	NA	7	15	22	22
Others	NA	NA	12	9	10	6

NA = Not available.

* = Less than 1 percent.

[1]Estimated for corn, cotton, potatoes, soybeans, and wheat. Since potatoes were not surveyed in 2000, the 2000 estimate includes potato use in 1999.

[2]Alachlor, acetochlor, metolachlor, propachlor.

[3]Oryzalin, pendimethalin, ethalfluralin, trifluralin.

[4]Butylate, EPTC, pebulate.

[5]2,4-D, 2,4-DB, MCPA, MCPB.

[6]Atrazine, cyanazine, propazine, simazine, metribuzin, ametryne.

[7]Glyphosate, glufosinate-ammonium, sulfosate.

[8]Chlorsulfuron, halosulfuron, metsulfuron, nicosulfuron, primisulfuron.

[9]Includes bipyridyls (paraquat), benzothiadiazoles (bentazon), benoxazoles (fenaxaprop), imidizolinones (imazaquin, imazethapyr), diphenyl ethers (acifluorfen, diclofop, lactofen, oxyfluorfen), oximes (clethodim, clomazone, sethoxydim), pyridines (clorpyralid, fluazifop), pyridazinones (norfluorazon), and others that first appeared in pesticide use surveys since 1976.

[10]Sum of acreage treated with a pesticide multiplied by average number of applications per acre.

Source: Eichers et al., 1968; Andrelenas, 1974; unpublished ERS data.

Among herbicides, shares of acre-treatments and quantity for phosphinic acids—primarily glyphosate (trade name: Roundup) but also glufosinate-ammonium and sulfosate—were much higher in 2000 than in 1996 (Table 21). At the same time, shares of amides, anilines, phenoxys, and triazines, widely used since the 1960s and 1970s, were lower in 2000. Shares of sulfonyl ureas and other new families increased before 1996, but there was little change between 1996 and 2000.

Phosphinic acid shares were higher in 2000 than in 1996 on all major crops, but especially cotton and soybeans, where adoption of genetically modified seed tolerant to these herbicides has been widespread (see Chapter 9, "Biotechnology and Agriculture"). Phosphinic acids were the most used herbicides on cotton and soybeans in 2000; their share of herbicide acre-treatments increased from 4 percent in 1996 to 30 percent in 2000 on cotton and from 10 to 42 percent on soybeans, while shares of other major herbicide families were stable or declined. The higher application rates with phosphinic acids contributed to the higher soybean herbicide quantity in 2002 relative to previous years (Table 19). The phosphinic acid share of herbicide acre-treatments increased from 2 percent in 1996 to 6 percent in 2000 on corn, 7 to 13 percent on winter wheat, 3 to 7 percent on durum wheat, and 4 to 10 percent on other spring wheat. However, shares of sulfonyl ureas and other new herbicide families on corn and wheat were higher in 2000 than in 1996. (see "Crop Production Practices" in the ARMS Data Tool on the ERS website.)

Since some producers used phosphinic acids and other newer post-emergence herbicides (applied after weed emergence) instead of older pre-emergence herbicides, the shares of corn, cotton, soybean, and spring wheat acres receiving post-emergence applications were higher and acres receiving pre-emergence applications lower in 2000 than in 1996. While producers treat many acres with both pre- and post-emergence herbicides, the shares of acres receiving post-emergence applications only were higher in 2000 than 1996, with the share of soybean acreage almost doubling from 28 percent in 1996 to 50 percent in 2000. (See "Crop Production Practices" in the ARMS Data Tool on the ERS website.)

PESTICIDE PRICES

The USDA/NASS agricultural chemical price index was relatively stable from 1996 through 2003, while the herbicide price index declined by 4 percent, the insecticide price index increased by 17 percent, and the fungicide/other index was stable (Table 22). Prices of individual pesticides may behave differently, responding to different factors. The price for glyphosate fell by 22 percent from 1996 to 2003, which may reflect marketing strategy as well as its patent's expiring in September 2000. The lower price may have encouraged producers to use glyphosate and genetically modified herbicide-tolerant seed. The price for methyl bromide increased by 147 percent from 1996 to 2003, because EPA required supply reductions to implement the Montreal Protocol phaseout, beginning in 1999. Higher prices encouraged producers to use other pesticides, and focused methyl bromide use on crops and acres with higher returns.

Based on USDA/NASS indices, pesticide prices have risen more slowly than wages, fuel prices, and crop prices since the late 1990s, which departs from the post-1980 trend of pesticide prices rising faster than crop and fuel prices (Figure 45, Table 22). Pesticide prices have risen more slowly than crop prices since 2000 and fuel prices since 1998. Recent trends could encourage producers to substitute pesticides for fuel- and labor-intensive practices, but the slowdown in pesticide prices could also reflect declining demand.

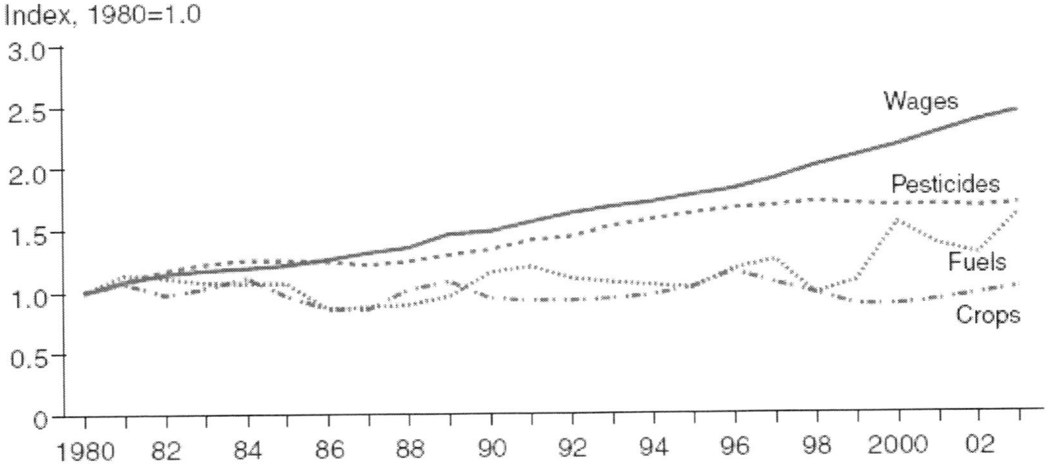

Source: NASS/USDA.

Figure 45. Price indices: crops, wages, fuels, and pesticides, 1980-2003.

Table 22. Price indices and selected pesticide prices, 1996-2003

Index or price	1996	1997	1998	1999	2000	2001	2002	2003
				(1990-92 = 100)				
Price indices								
Agricultural chemicals	119	121	122	121	120	121	119	121
Herbicides	117	117	118	114	111	112	111	112
Insecticides	125	130	136	141	145	144	140	146
Fungicides and others	117	119	119	120	120	120	121	117
Fuels	102	106	84	93	134	119	112	140
Wage rates	117	123	129	135	140	146	153	157
Crops	127	115	107	97	96	99	105	111
					($/lb. a.i.)			
Selected pesticide prices								
Glyphosate	13.93	14.18	14.08	11.38	10.83	11.13	10.88	10.88
Methyl bromide	3.02	3.31	3.23	3.15	3.58	4.97	5.42	7.45

Source: NASS/USDA.

PEST MANAGEMENT PRACTICES

Growers use biological and cultural practices and information to improve the cost effectiveness of pest management, often coordinating their use through Integrated Pest Management (IPM). Use of practices varies by crop because of different pests and production requirements. Cotton growers use many practices, especially insecticide applications, more intensively than do growers of other crops because of the crop's high value and its vulnerability to pests (especially insects) (See "Crop Production Practices" in the ARMS Data Tool on the ERS website.)

Corn, cotton, soybean, and wheat growers reported tilling, chopping, mowing, or cleaning equipment for pest control on more than 30 percent of acres in 2000, but cotton growers reported the highest proportions. Cultivation for weed control was higher in 1996 than in 2000 on cotton (89 percent of acres versus 65 percent) and soybeans (30 percent versus 17 percent). This may reflect increased use of genetically modified herbicide-tolerant seed and post-emergence herbicides. The share of corn acres cultivated for weed control was higher in 2000 (38 percent) than in 1996 (32 percent).

Among other practices, growers reported adjusting planting or harvest dates to manage pests on about 20 percent of cotton and wheat acres in 2000–more than on corn or soybeans. Growers reported alternating pesticides to prevent pest resistance on 30 percent or more of corn, cotton, soybean, and spring wheat acres in 1996 and 2000, but only about 10 percent of winter wheat acres. (Pesticides are generally applied to 50 percent or less of winter wheat, but to 90 percent or more of the other crop acres.) Cotton growers reported protecting beneficial organisms on the highest proportion of acres, approximately 50 percent in 1996 and 2000.

Cotton growers reported more scouting for insects and reliance on independent consultants or scouts than did growers of other crops in 1996 and 2000 (See Chapter 18, "Information Technology Management"). Corn, cotton, soybean, and wheat growers reported scouting for weeds, insects, and diseases on 50 percent of acres or more. With the exception of scouting for insects on cotton, operators, partners, or family members scouted the most acreage for insects and weeds, more so than farm supply/chemical dealers and independent consultants and scouts. Independent consultants or scouts had the largest role on cotton, scouting for insects on about 50 percent of cotton acres and weeds on 20-25 percent. On corn, they scouted about 10 percent of acres for insects and 12 percent for weeds.

Farm supply or chemical dealers were identified by growers as primary pest management information sources on corn, soybean, and wheat acreage, ranging from 40 percent on wheat to over 60 percent on corn and soybeans in 1996 and 2000. Cotton growers relied more on independent crop consultants or pest control advisors (30 percent of acres in 2000), Extension (17 percent), and commercial scouting (10 percent) than did growers of other crops, with farm supply or chemical dealers (26 percent) the second most identified source. Winter wheat producers reported no pest management information source for 22 percent of acres in 2000, which may reflect their less intensive use of pesticides.

POLICY AND REGULATORY ISSUES

Asian Soybean Rust

Asian soybean rust is caused by a windborne, highly prolific, and virulent fungal pathogen (*Phakopsora pachyrhizi*) that can infect over 95 species of cultivated and wild plants, including soybeans and kudzu. The pathogen has caused yield losses and higher production costs in Asia, Australia, Africa, India, and South America. Responding to its introduction in South America, USDA's Animal and Plant Health Inspection Service established a rust surveillance, information, and education program in 2002 to help domestic producers respond effectively. Asian soybean rust was first identified in the United States in late 2004.

Historically, producers treat less than 1 percent of U.S. soybean acres with fungicides (excluding seed treatments). To prevent production losses from soybean rust, U.S. producers might increase fungicide use by 2.5-10.5 million pounds a.i. per year, with 20 to over 90 percent of soybean acres treated, depending upon the severity and extent of outbreak. This would increase production costs (Livingston et al., 2004). Total fungicide use on crops could increase 7 to 30 percent over the 34 million pounds a.i. estimated in 2002. Fungicides registered for soybean rust include azoxystrobin, chlorothalonil, and pyraclostrobin. In addition, EPA granted Federal Insecticide Fungicide and Rodenticide Act emergency exemptions for propicona-zole, tebuconazole, myclobutanil, tetraconazole, and the combination of trifloxystrobin plus propiconazole. Some have been used successfully in other countries.

Scientists estimate that soybean rust could reduce yields of untreated soybeans by 10 to 60 percent. Based on evidence that fungicide use would limit average losses to about 4 percent, Livingston et al. (2004) estimated U.S. producer and consumer losses from soybean rust could vary between $240 million and $2 billion per year over 5 years, depending upon the extent and severity of an outbreak.

Food Quality Protection Act

The potential dietary, drinking water, worker, human health, and environmental hazards of pesticide use often are not completely reflected in producers'costs and returns. So the Federal Insecticide, Fungicide and Rodenticide Act of 1947 (FIFRA) regulates which pesticides can be used on crops and how they can be used, through EPA's pesticide registration process. The Federal Food, Drug, and Cosmetic Act of 1938 (FFDCA) regulates pesticide residues in food.

The Food Quality Protection Act of 1996 (FQPA) amended FIFRA and FFDCA to set new standards for and to modify the regulation of pesticide residues in food. Under FQPA, EPA must consider dietary exposure from all food uses and drinking water, nonoccupational exposure such as homeowner use, and the susceptibility of infants and children in setting pesticide residue tolerances, as well as the cumulative effects of substances if there is a common mechanism of toxicity. FQPA required a reassessment of all existing pesticide residue tolerances by 2006, with priority to pesticides that pose the greatest risk to public health. EPA is coordinating the tolerance reassessment with the reregistration of pesticides to comply with new standards mandated in amendments to FIFRA in 1988.

The reassessment resulted in revocations or modifications of some residue tolerances and cancellations or restrictions of some use registrations. Among the highest priorities are pesticides in the carbamate, organochlorine, and organophosphate families, or pesticides classified as carcinogens. EPA met FQPA-mandated interim goals, and by the end of fiscal year 2004 had reassessed about 73 percent of the 9,721 mandated tolerances, including about 67 percent of 1,691 organophosphate, 57 percent of 545 carbamate, 71 percent of 2,008 carcinogen, and all 253 organochlorine tolerances (USEPA, 2005a). Many reassessed organophosphate tolerances required no modification, but EPA restricted or cancelled use of azinphos methyl, chlor-pyrifos, and methyl parathion on some crops due to dietary risk. EPA cancelled use of chlopyrifos by homeowners and in schools, parks, and other settings, as well as outdoor residential use of diazinon to reduce risks to children. EPA is conducting a cumulative assessment of organophosphate tolerances that could lead to further actions.

Methyl Bromide Phaseout

Methyl bromide is used for soil fumigation before planting many fruit and vegetable crops, post-harvest storage and facility fumigation, and government-required quarantine treatments. It was identified as an ozone-depleting substance under the Montreal Protocol, implemented in the United States through the Clean Air Act. Its use was incrementally phased out in developed countries from 25 percent of the 1991 use baseline beginning January 1, 1999, to 100 percent on January 1, 2005. Its use will be phased out by 2015 in developing countries. The Protocol's Quarantine and Preshipment (QPS) and Critical Use Exemptions allow some methyl bromide use in developed countries after the phaseout. QPS treatments are permitted to meet some government phytosanitary and quarantine requirements for imports and exports, and some standards of Federal, State, and local governments.

The Parties to the Montreal Protocol can grant critical-use exemptions for specific uses in a country if no technically and economically feasible alternative with acceptable health and environmental effects is available, and if a significant market disruption would occur without methyl bromide, but the country must take steps to develop alternatives and minimize methyl bromide use and emissions (Osteen, 2003). Countries requesting exemptions submit annual nominations, and the approval process has been contentious. The United States requested more methyl bromide for 2005 and 2006 than permitted under the 2003 reduction goal—30 percent of its 1991 baseline of 56.3 million pounds. The Parties approved quantities for the U.S. in 2005 totaling 37 percent of its baseline; however, permitted production and imports would satisfy only 30 percent, with the remainder coming from existing U.S. stockpiles (USEPA, 2005b). For 2006, the U.S. requested exemptions totaling 37 percent of the baseline, and the Parties approved quantities totaling 32 percent. For 2007, the United States requested 29 percent of the baseline.

References

[1] Andrelenas, P.A. (1974). Farmers' Use of Pesticides in 1971 – Quantities. AER-252, U.S. Dept. Agr., Econ. Res. Serv.

[2] Eichers, T.R., P.A. Andrelenas, R. Jenkins, and A. Fox (1968). Quantities of Pesticides Used by Farmers in 1964. AER-131, U.S. Dept. Agr., Econ. Res. Serv.

[3] Kiely, T., D. Donaldson, and A. Grube (2004). Pesticide Industry Sales and Usage: 2000 and 2001 Market Estimates, Biological and Economic Analysis Division, Office of Pesticide Programs, Office of Prevention, Pesticides, and Toxic Substances, U.S. Environmental Protection Agency, May.

[4] Livingston, M.J., R. Johansson, S. Daberkow, M. Roberts, M. Ash, and V. Breneman (2004). Economic and Policy Implications of Wind-Borne Entry of Asian Soybean Rust into the United States. OCS–04D–02, U.S. Dept. Agr., Econ. Res. Serv., April.

[5] Osteen, Craig (2003). "Methyl Bromide Phaseout Proceeds: Users Request Exemptions," Amber Waves, vol. 1, no. 2, April.

[6] Padgitt, M., D. Newton, and C. Sandretto (2000). *Production Practices for Major Crops in U.S. Agriculture, 1990-97.* SB-969, U.S. Dept. Agr., Econ. Res. Serv.

[7] U.S. Environmental Protection Agency, Office of Pesticide Programs (2000). "Malathion Summary."

[8] U.S. Environmental Protection Agency, Office of Pesticide Programs (2005a). *Taking Care of Business: Protecting Public Health and the Environment, EPA's Pesticide Program FY2004 Annual Report*, EPA-735-R-05-001.

[9] U.S Environmental Protection Agency, Office of Air and Radiation (2005b). "The Phaseout of Methyl Bromide," available at www.epa.gov/ozone/mbr

In: Agricultural Resources and Environmental Indicators ISBN 978-1-60021-467-7
Editor: Keith Wiebe and Noel Gollehon, pp. 139-145

Chapter 15

NUTRIENT MANAGEMENT

Stan Daberkow and Wen Huang

Since the early 1990s, U.S. commercial fertilizer use, application rates, and management practices have tended to change modestly from year to year, while fertilizer prices have exhibited more variability.

INTRODUCTION

The major plant nutrients (nitrogen, phosphorus, and potassium) are critical for maintaining crop yields but have also been associated with the impairment of numerous streams, lakes, and aquifers. For most U.S. crops and in most regions, commercial fertilizer is the major source of plant nutrients, although organic sources—such as legumes, crop residue, and animal wastes—can also provide nutrients required for plant growth. Commercial fertilizer is a major agricultural input; farmers typically spend over $10 billion annually on commercial fertilizer, although fertilizer use and prices vary from year to year. Historically, crop producers have used large amounts of commercial fertilizer and organic nutrients, but the concern over runoff and leaching has prompted the promotion of nutrient management practices that minimize nutrient loss.

The share of acres receiving fertilizer, application rates for primary nutrients (nitrogen, phosphate, and potash), and nutrient management practices on major field crops (corn, soybeans, wheat, and cotton) remained fairly stable over the 1990s. However, fertilizer prices, especially for nitrogen, have been volatile and have risen rapidly in recent years. Despite increased fertilizer prices and growing concern about environmental risks from fertilizer use, the use of nutrient management practices on major crops has changed little since the early 1990s.

FERTILIZER USE NATIONALLY AND BY REGION

U.S. commercial fertilizer use peaked in 1981 at over 23 million nutrient tons, but has exceeded 22 million tons seldom since then (Figure 46). The decline in (principal crop) planted acreage since 1998 likely accounts for part of the falloff in fertilizer since then. The mix of crops planted each year also influences aggregate fertilizer use. Corn and wheat acreage, which consumes the most fertilizer among all crops, has dropped since 1998, and fertilizer use on soybeans has only partially offset the falloff.

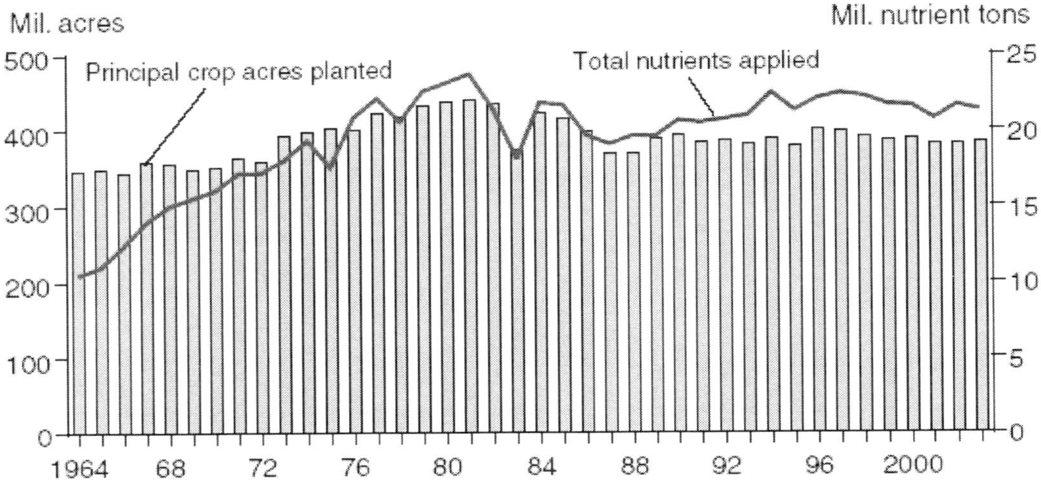

Source: USDA, Economic Research Service.

Figure 46. Principal crop acres planted and total nutrient use, 1964-2003.

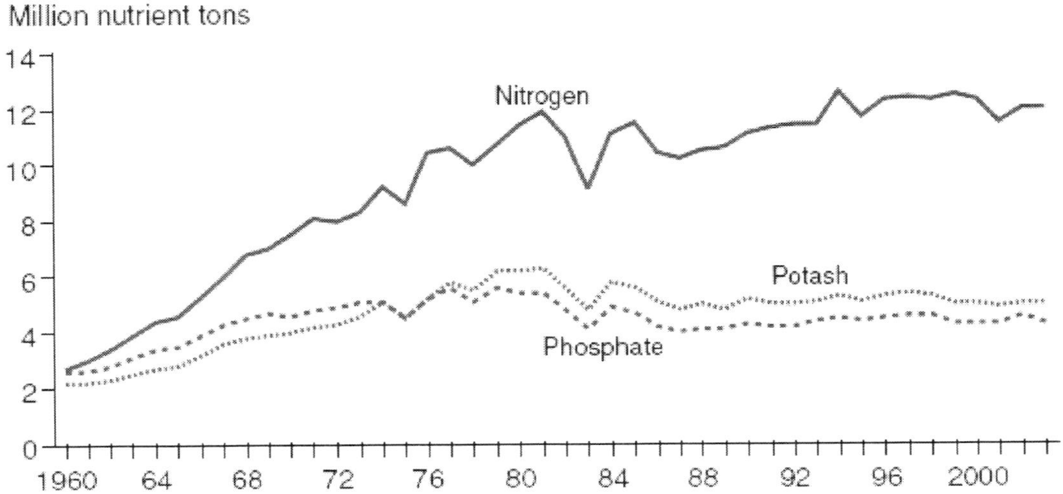

Source: USDA, Economic Research Service.

Figure 47. Consumption of primary plant nutrients, 1960-2003.

Consumption of individual nutrients has been variable over the last several years, although annual use through 2003 is below levels reported in the late 1990s (Figure 47). For example, nitrogen use dropped noticeably in 2001 and remained below 12.5 million tons from 2000 through 2003. Annual phosphate and potash use has demonstrated similar variability over the last several years. The regional distribution of fertilizer use has remained stable, with the Corn Belt, Northern Plains, and Lake States the leading regions because of high concentrations of corn, wheat, and soybean acreage (Figure 48).

FERTILIZER USE BY MAJOR CROPS

The four major U.S. crops—corn, wheat, soybeans, and cotton—currently account for about 60 percent of principal crop acreage and receive over 60 percent of all nitrogen, phosphate, and potash used in the United States. Corn typically accounts for over 40 percent of all commercial fertilizer consumed, followed by wheat (about 10 percent), soybeans (about 5 percent), and cotton (5 percent). However, these shares vary from year to year (and by nutrient) due to the mix of crops planted, share of acreage treated, and application rates.

The share of acreage treated and application rates were fairly stable over 1990-2003, although cotton exhibited greater variability (Table 23). For example, 97-98 percent of corn acres received nitrogen fertilizer in most years (excluding the high and low years), while phosphate was applied to 79-84 percent of the acres. Similarly, for most crop/nutrient combinations, the amount of nutrients applied each year varied 5 pounds or less. This relative consistency in production practices stems from modest changes in factors like the ratio of prices received to fertilizer prices paid by farmers, agronomic relationships, seed traits, public policies, and producer education.

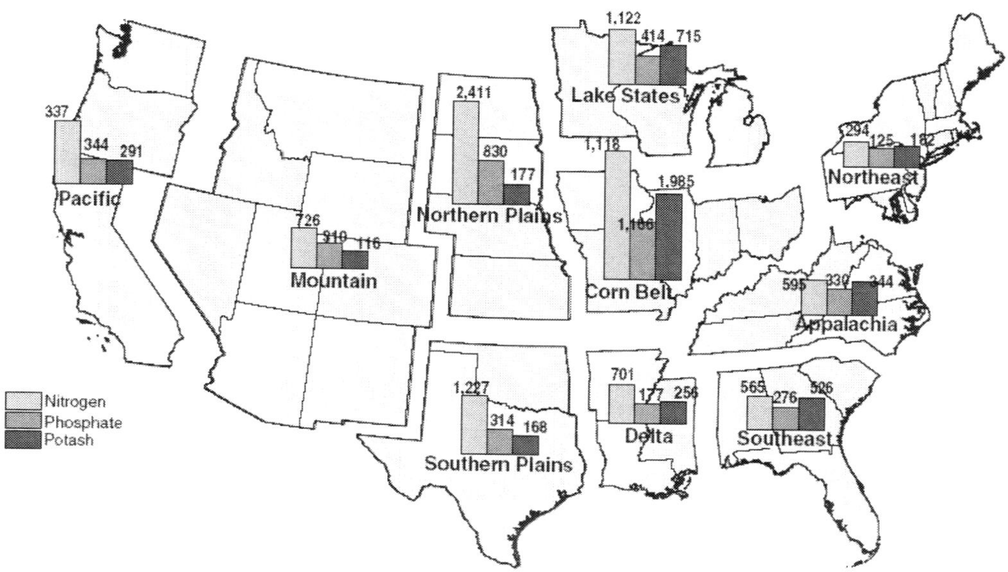

Source: Association of American Plant Food Control Officials.

Figure 48. Fertilizer consumption by farm production region, year ending June 30, 2003.

Table 23. Share of acres treated and application rates, 1990-2003, by major crop and nutrient[1]

	Nitrogen	Phosphate	Potash
Corn			
Share of acres treated (%)	97-98	79-84	65-73
Application rates (lbs/treated acre)	127-136	56-59	80-84
Cotton			
Share of acres treated (%)	79-86	49-66	34-53
Application rates (lbs/treated acre)	84-100	44-49	48-76
Soybean			
Share of acres treated (%)	13-18	21-26	25-29
Application rates (lbs/treated acre)	22-25	47-50	76-88
Wheat			
Share of acres treated (%)	80-88	54-63	18-20
Application rates (lbs/treated acre)	63-68	32-35	37-41

[1]Excludes values for high and low years which may have been influenced by such factors as number of States surveyed, weather, commodity and/or fertilizer prices, etc.
Source: ERS from NASS data on Agricultural Chemical Usage

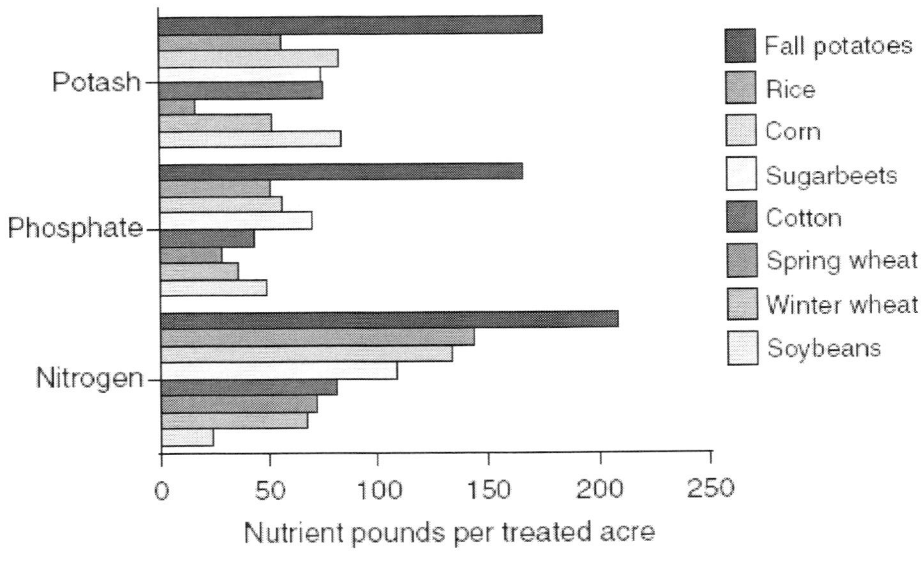

[1]Data for wheat, rice and sugarbeets are for 2000.
Source: USDA, Economic Research Service.

Figure 49. Average application rates of commercial fertilizers, by selected crops, 20011.

While nutrient use, in general, has been stable since 1990, modest trends for some crop/nutrient combinations are apparent. For example, the share of corn acres treated with phosphate and potash declined slightly, while the share of cotton acres treated with potash rose along with application rates. The share of wheat acres treated with nitrogen likewise rose, as did application rates; these increases may have been due to the decline of wheat acreage in several arid States in the Mountain and Plains regions. Also, a significant price increase for fertilizer, especially nitrogen, in 2001 and 2003 likely dampened nutrient use during those years.

Fertilizer application rates vary widely by crop and nutrient, and are influenced by yield response, climate, and fertilizer/commodity prices (Figure 49). Among the major field crops, annual application rates for all nutrients are typically highest for corn, but rates vary widely for other crops and nutrients. Certain specialty crops, like fall potatoes and rice, consume more fertilizer per acre than the major crops, but are planted on far fewer acres.

PRICES OF MAJOR FERTILIZER PRODUCTS

Since 1990, the primary fertilizer products purchased by farmers have exhibited much different nominal price patterns (Figure 50). Potash prices, in the form of potassium chloride (KCl), have been stable. A major source of phosphate, diammonium phosphate (DAP), has demonstrated much more price variability—peaking in 1995 at near-record levels, declining steadily through 2002, then rising again in 2003 and 2004. Nitrogen product prices, such as those for anhydrous ammonia, have shown the greatest volatility during the last several years—a spike in 1995, a fall to historically low levels by 1999 and 2000, record-high prices in 2001, followed by a large decline in 2002 and recovery in 2003. Fertilizer markets are influenced by trade, raw material prices, and planted acres. Potash is largely an imported product, whereas the United States exports large amounts of phosphate; both of these markets are influenced by international demand and supply factors. Natural gas is the primary raw material for nitrogen fertilizers, comprising 75 percent or more of their cost of production. During 2001 and 2003, nitrogen fertilizer prices rose dramatically in concert with natural gas prices. In response to higher prices in the United States, nitrogen-based fertilizer imports have increased significantly in recent years.

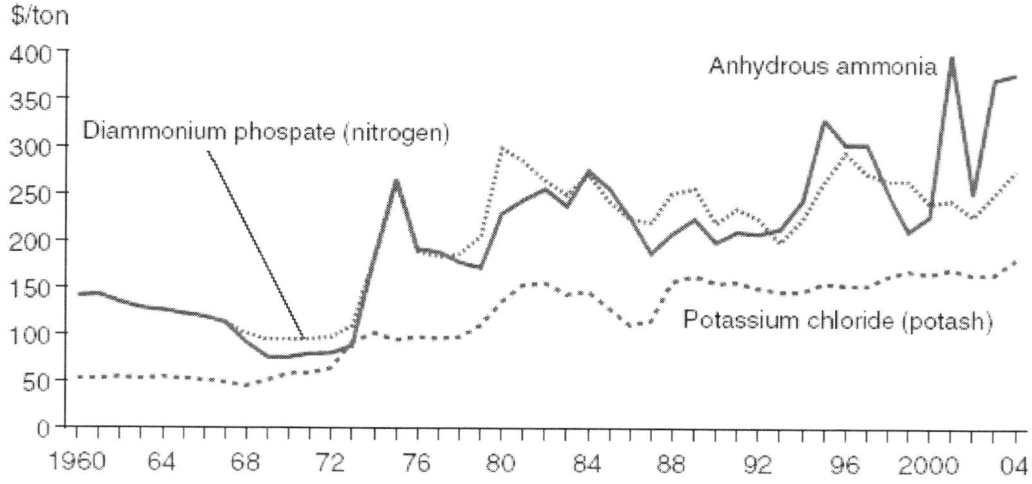

Source: USDA, Economic Research Service.

Figure 50. Farm prices for anhydrous ammonia, DAP, and KCl, 1960-2004.

While most economic studies indicate that farmer response to fertilizer price changes is fairly inelastic (i.e., relatively unresponsive to price changes), farmers can make some adjustments in fertilizer use when faced with dramatically higher prices. Shifting to a crop that needs less fertilizer, reducing application rates, and adopting improved nutrient management practices are all options. When the 2001 ARMS asked corn producers how they responded to higher nitrogen fertilizer prices, about one-third reported that most of their nitrogen had been contracted at a pre-determined price, so they were not affected by increased fertilizer prices. Another 11 percent reported reducing application rates or changing nutrient management practices. The remaining producers, who did not change their nutrient use or management, tended to operate smaller farms and use less nitrogen per acre.

Nutrient Management Practices for Major Crops

The use of nutrient management practices in crop production can have economic and environmental implications (Heimlich, 2003; and "Agricultural Chemicals and Production Technology" briefing room on the ERS website). For example, using soil tests to assess the need for additional commercial fertilizer or manure applications can reduce fertilizer costs and losses to the environment. Applying nitrogen inhibitors to delay the release of nitrates from ammonium fertilizers until later in the growing season may reduce nitrate leaching. If nitrogen fertilizer products are broadcast, incorporating the product into the soil may reduce nitrogen losses through volatilization. Applying nitrogen at strategic times (i.e., split applications), such as after planting when crop demand is greatest, may reduce the risk of nitrogen loss through leaching or volatization.

In general, ARMS data indicate that the use of most nutrient management practices has remained steady from year to year for the major crops. Between 1996 and 2000, soil testing was conducted on 40-50 percent of the acres planted to corn and cotton—the crops with the largest fertilizer use per acre. On soybeans and wheat, only about 30 percent of the planted acreage was soil tested. No clear soil-testing trends were reported for any of these crops. Similarly, with the possible exception of winter wheat and corn, nitrogen management practices, such as fall application, split application, and incorporation of broadcast materials, showed little consistent change over time. The share of corn acres with all nitrogen broadcasted without incorporation declined from about 15 percent to 9 percent between 1996 and 2000. The share of winter wheat acres with split nitrogen application increased, as did non-incorporated broadcast acres. Nitrogen inhibitors were used on 7-9 percent of U.S. corn acreage from 1996 to 2000, but were used on less than 2 percent of other major crops. As with the share of acres treated and application rates, nutrient management practices tend to change little from year to year.

Programs and Regulations That May Affect Fertilizer Use and Management

Nitrogen and phosphorus have been identified as major contaminants of U.S. surface and ground water (see Chapter 5, "Water Quality: Impacts of Agriculture"). The U.S. Geological Survey (1999) estimates that about 90 percent of nitrogen and 75 percent of phosphorus contaminants originate from nonpoint sources, with the remainder from point sources. Agricultural point sources include livestock operations, while nonpoint sources include fertilizers and animal waste applied to cropland.

Given the concerns over water quality, a number of voluntary and nonvoluntary programs have been promulgated to address agricultural nutrient use and management. Most Federal programs are directed at encouraging producers to alter cropping or nutrient application practices. These programs range from nutrient use regulations affecting producers of large confined animal operations to voluntary cost-sharing and educational/technical assistance programs available to all producers. Livestock producers who meet the criteria for a concentrated animal feeding operation (CAFO) must formulate a nutrient management plan for animal waste disposal that includes record-keeping, and in certain cases, limits on the application of other sources of nutrients, such as commercial fertilizer. The Environmental Quality Incentives Program (EQIP) and the Conservation Security Program (CSP) provide cost-sharing to producers who adopt environmentally friendly practices, including a nutrient management plan. Such plans focus on managing the amount, source, placement, and timing of fertilizers and animal manure to minimize pollution on cropland. Other Federal programs include Conservation Compliance, which reduces nutrient losses associated with soil erosion, the Conservation Reserve Program (CRP), which has retired over 34 million acres of environmentally sensitive cropland under 10- to 15-year contracts, and Conservation Technical Assistance (CTA). More recently, the Conservation Reserve Enhancement Program (CREP) and the Wetland Reserve Program (WRP) have been implemented to remove environmentally sensitive land from crop production (i.e., buffers, filter strips, and wetlands). (See Part 5.)

References

[1] Association of American Plant Food Control Officials (2003). *Commercial Fertilizers*, University of Kentucky, Lexington, KY.

[2] Heimlich, R. (2003). *Agricultural Resources and Environmental Indicators, 2003*. AH-722, U.S. Dept. Agr., Econ. Res. Serv.

[3] U.S. Department of Agriculture, Economic Research Service (2005) "U.S. Fertilizer Use and Price." Data product on the ERS website.

[4] U.S. Department of Agriculture, National Agriculture Statistics Service (various years) *Agricultural Chemical Usage, Field Crop Summary.*

[5] U.S. Department of Agriculture, National Agricultural Statistics Service (various years) *Agricultural Prices—Annual Summary.*

[6] U.S. Department of Agriculture, National Agricultural Statistics Service (various years) *Crop Production—Annual Summary.*

[7] U.S. Geological Survey (1999). *The Quality of Our Nation's Waters: Nutrients and Pesticides.* U.S. Geological Survey Circular 1225.

In: Agricultural Resources and Environmental Indicators ISBN 978-1-60021-467-7
Editor: Keith Wiebe and Noel Gollehon, pp. 147-156

Chapter 16

ANIMAL AGRICULTURE AND THE ENVIRONMENT

Marc Ribaudo and Noel Gollehon

Animal production produces a number of pollutants that can affect the quality of air and water resources. Environmental/conservation policies and programs provide incentives to farmers to reduce pollution from animal operations.

INTRODUCTION

Animal production industries have seen substantial changes over the past several decades, the result of domestic/export market forces and technological changes. The number of large operations has increased, and animal and feed production are increasingly separated in terms of both management and geography. Concern that these changes are harming the environment has prompted local, State, and Federal policies (see Chapter 27, "Federal Laws Protecting Environmental Quality") and programs to control pollution from animal production facilities.

TRENDS IN ANIMAL PRODUCTION AND MANURE NUTRIENTS

Changes in the structure of livestock and poultry production are behind many of the current concerns about animals and the environment. Structural changes have been driven by both innovation and economies of size (McBride and Key, 2003). Organizational innovations, such as production contract arrangements, enable growers to access the capital necessary to adopt innovative technologies and garner economies of size, with greater profit potential. The significant economic benefits from vertical coordination, particularly for poultry and swine operations, have led to both larger operations and greater geographic concentration of animals.

The number of U.S. farms with confined animals (called animal feeding operations, or AFOs) has declined steadily from 435,000 in 1982 to 213,000 in 1997 (Gollehon et al., 2001). Declines occurred in all sectors, but primarily in the very small and small farm sizes (see box, "Size Groupings"). This decline in farms has been accompanied by a 10-percent increase in the number of confined animal units (AUs, defined as 1,000 pounds of live weight) (Figure 51). A decline in AUs on very small and small farms was more than offset by growth on medium-sized farms and large farms (Gollehon et al., 2001).

The regional distribution of confined animals also changed between 1982 and 1997. Animal populations in the Prairie Gateway and Southern Seaboard regions increased by 2 million (40 percent) and 1.7 million (70 percent) animal units over 1982-97 (Figure 52). Only the Northern Crescent and Heartland regions exhibited significant declines.

Size Groupings

One animal unit is defined as 1,000 lbs live weight (e.g. 1 AU = 1.14 feedlot beef, 0.74 dairy cow, 9.09 swine for slaughter, or 455 broilers).

Animal operations are classified as:

Very small, less than 50 AU

Small, 50-299 AU

Medium, 300-999 AU

Large, more than 1,000 AU

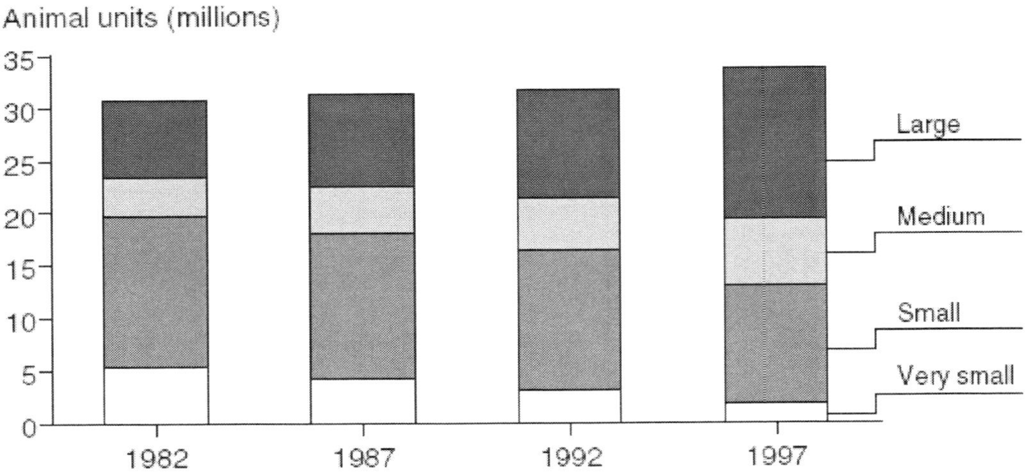

Source: Census of Agriculture and ERS.

Figure 51. Confined animal units by size of animal operation, 1982-97.

The innovation and economies of size that underlie changes in the livestock and poultry sector also served to separate animal production from crop production. Large, specialized facilities today focus on producing animals and purchase most of their feed from off the farm.

This means there is generally less land on the animal farm on which to spread manure. The amount of land per animal unit declined nearly 40 percent across all animal types between 1982 and 1997, from 3.6 to 2.2 acres per AU. (See Gollehon et al, 2001, for additional information on trends).

ENVIRONMENTAL IMPACTS OF ANIMAL PRODUCTION

The major source of environmental degradation from confined animal production is the wastes (manure, urine, bedding material) that are produced. Animal waste can be transmitted through runoff of nutrients, organic matter, and pathogens to surface water; leaching of nitrogen and pathogens to ground water; and volatilization of gases and odors to the atmosphere. Pollutants may originate at production houses/lots where animals are kept; manure storage structures such as tanks, ponds, and lagoons; or land where manure collects or is applied.

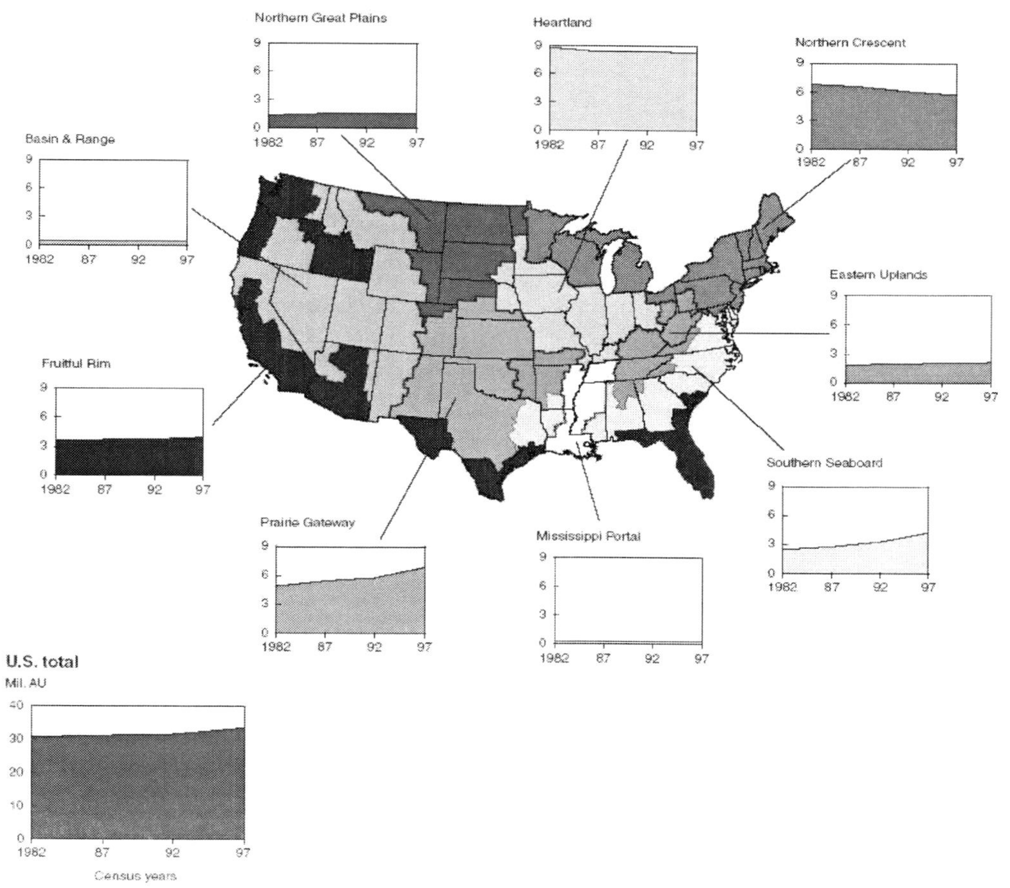

Source: Economic Research Service, USDA.

Figure 52. Confined animal units by ERS Resource Region, 1982-97.

The major pollutants include:

- **Nutrients**—Nitrogen and phosphorus are essential plant nutrients, but can degrade water quality by causing eutrophication (see Chapter 5, "Water Quality: Impacts of Agriculture).
- **Ammonia**—A pungent, colorless gas that can be a health hazard to humans and animals at high concentrations, and a precursor for fine particulates (haze) in the atmosphere. It also contributes to soil acidification and eutrophication.
- **Hydrogen sulfide**—A colorless gas also hazardous to humans and animals.
- **Methane**—A nontoxic, odorless gas that contributes to global warming (greenhouse gas).
- **Odor**—A nuisance associated with animal production facilities. Odorous gases consist of a host of compounds (over 160) that originate from manure in animal housing, manure storage units, and land application.
- **Pathogens**—Threats to human health that are often contained in manure. Some of the pathogens that pose a threat to human health include the protozoan parasites *Cryptosporidium* and *Giardia* and some bacteria species such as *Salmonella*, *E. coli,* andCampylobacter.

Manure Production and Excess Nutrients

Two indicators of potential environmental degradation from animal feeding operations are total nitrogen excreted and excess nitrogen and phosphorus. Total nitrogen is an indicator of the potential for both air and water pollution from the entire operation (production facility, manure storage, and land application). Excess nutrients are manure nutrients produced on the farm in excess of the farm's crop needs. Excess nutrients are susceptible to running or leaching off the field and into water resources unless steps are taken to move the manure off the farm to additional land or to other industrial uses such as energy production or commercial fertilizer production.

In 1997, animal feeding operations controlled 73 million acres of cropland and permanent pasture. This land was estimated by Gollehon et al. (2001) to have the capacity to assimilate only 40 percent of the nitrogen and 30 percent of the phosphorus in the manure recoverable from animal production facilities and available as a crop fertilizer. Large farms, which constitute 2 percent of the total number of farms, accounted for almost half of the excess on-farm nutrients.

In 1997, 68 counties had manure nitrogen levels that exceeded the assimilative capacity of the entire county's crop and pasture land (Figure 53). Many more counties (152) have surplus manure phosphorus (Figure 54).

In these areas, it may be difficult to find enough land locally to spread manure without posing a risk to water quality. Research suggests that producers may have to haul manure extended distances in order to apply manure to land at agronomic rates (Ribaudo et al., 2003).

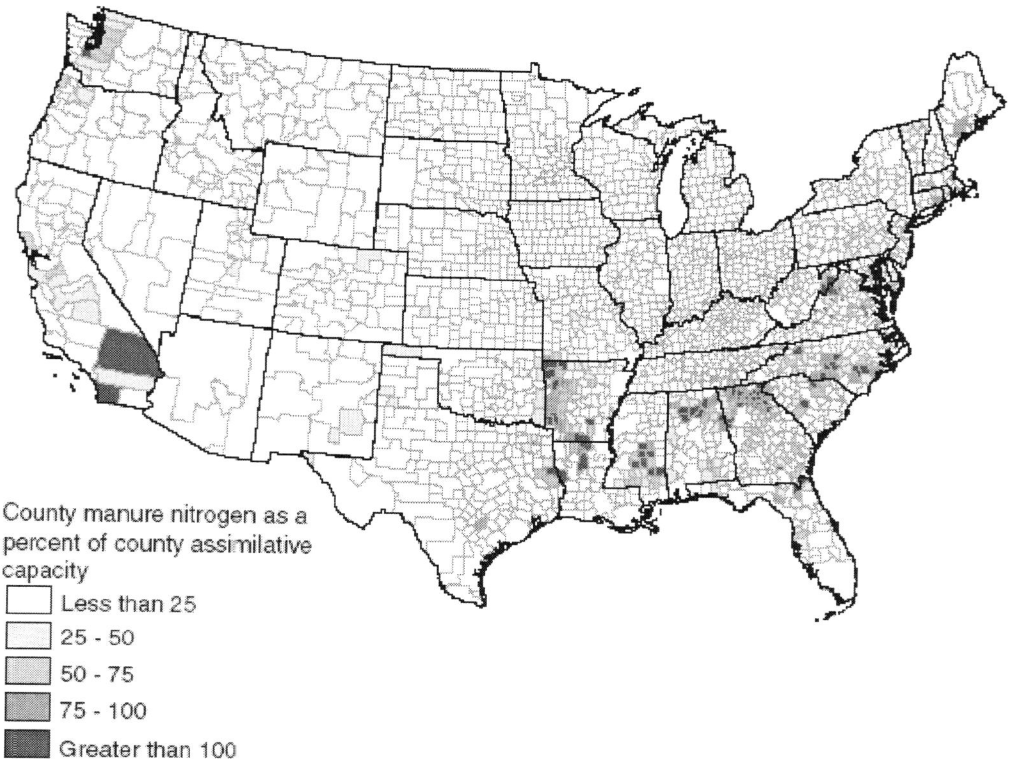

Some counties are combined to meet disclosure criteria.
Source: Economic Research Service, USDA.

Figure 53. Excess manure nitrogen as a share of county assimilative capacity, 1997.

MANURE'S CONTRIBUTION TO ENVIRONMENTAL DEGRADATION

While a nationwide study has yet to be completed, a number of studies have indicated that animal operations are significant contributors to water quality impairments in several regions. States reported to the Environmental Protection Agency (EPA) in 1996 that animal operations (feedlots, animal feeding operations, and animal holding areas) were a major factor in 5 percent of impaired rivers and streams, and a contributing source in 20 percent of rivers and streams reported as being impaired (U.S. EPA, 1998).[1] A United States Geological Survey (USGS) study of nitrogen loadings in 16 watersheds found that manure was the largest source in 6, primarily in the Southeast and Mid-Atlantic States (Puckett, 1994). In the Mississippi Basin, animal manure was estimated to contribute 15 percent of the nitrogen load entering the Gulf of Mexico; nitrogen is the suspected cause of a large zone of hypoxic waters (Goolsby et al., 1999). Monitoring by USGS in the National Water Quality Assessment Program found that the highest concentrations of nitrogen in streams occurred in agricultural basins, and were correlated with nitrogen inputs from fertilizers and manure (USGS, 1999). An analysis of fecal coliform bacteria in streams found that concentrations were partly a function of the number of both confined and unconfined animals in a watershed (Smith et al., 2005).

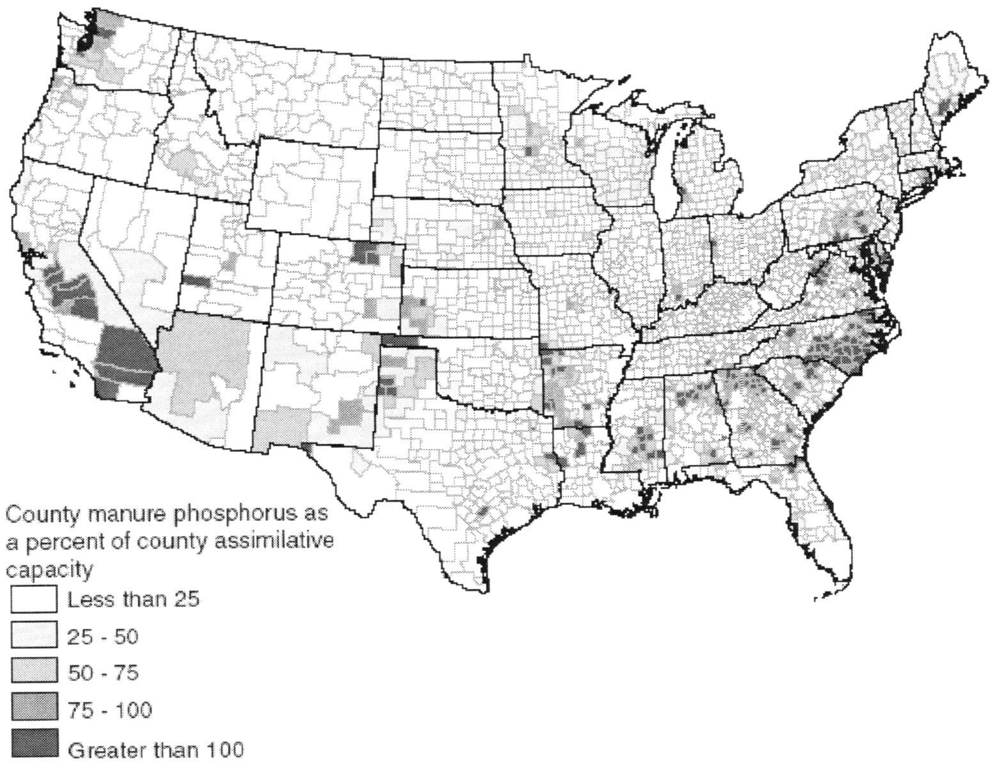

Some counties are combined to meet disclosure criteria.
Source: Economic Research Service, USDA.

Figure 54. Excess manure phosphorus as a share of county assimilative capacity, 1997.

The impact of gases and odor from animal feeding operations on human health or the environment has been difficult to determine because data on emissions are generally lacking (Jacobson et al., 1999). Animal waste in the United States has been estimated to contribute about 80 percent of all anthropogenic ammonia emissions, 25 percent of nitrous oxide emissions, and 18 percent of methane emissions (Battye et al., 1994; van Aardenne et al., 2001).

WATER-AIR INTERACTIONS

Emissions to water and to the atmosphere are not independent events, but are linked by the biological and chemical processes that produce the various compounds. For example, nitrogen excreted from an animal can follow any number of pathways and enter water as nitrate or the atmosphere as ammonia, nitrous oxide, nitric oxide, or as part of a volatile

[1] U.S. EPA's assessment relies on State self-reporting, which is incomplete and inconsistent between States (U.S. GAO, 2000). The Clean Water Act requires that such a report be submitted to Congress every 2 years.

organic compound. Reducing nitrogen movement along one pathway by changing its form will increase nitrogen movement along a different path (Figure 55). For example, reducing ammonia losses from a field by injecting waste directly into the soil increases the amount of nitrogen available for crop production, but may increase the risk of nitrate entering surface and ground water and nitrous oxide entering the atmosphere. The efficiency of manure management will depend on how these interactions are addressed. (For more information on this, see Aillery et al, 2005.)

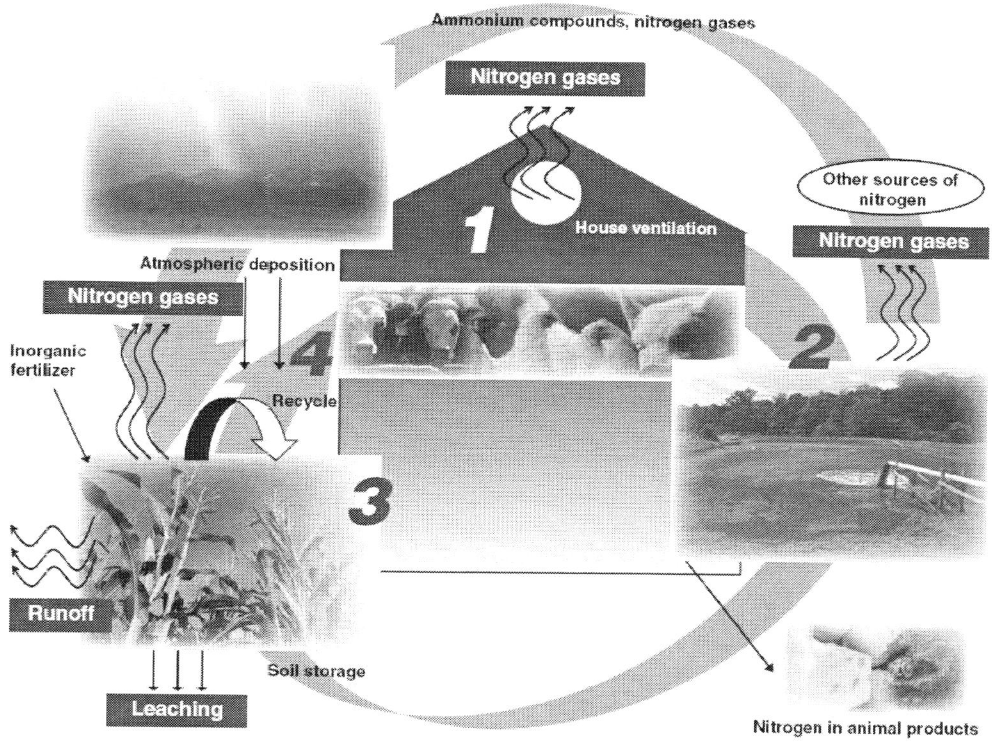

Source: Economic Research Service, USDA.

Figure 55. Nitrogen follows many pathways in a livestock operation.

Reducing Pollutant Losses

A number of practices are available for reducing gaseous emissions and runoff/leaching from animal feeding operations.

- ***Diet manipulation***—Feed additives and more efficient nutrient utilization in animals can reduce the amount of nitrogen and phosphorus in manure. This helps reduce the odor and ammonia emissions from production houses, and simplifies manure management for protecting water quality at all stages of handling and disposal.
- ***Chemical additive***—Different chemicals can be added to manure during collection in order to bind nutrients, thus reducing odorous compounds and ammonia emissions.

By reducing atmospheric emissions, the nitrogen content of manure increases, increasing its value as a fertilizer. But the higher nitrogen content can also increase the cost of applying manure at agronomic rates to protect water quality.

- ***Air treatment***—Trapping air vented from production houses and treating it before discharge to the atmosphere can reduce odorous compounds, ammonia, and other gases.
- ***Tank and lagoon cover***—Covering storage tanks and lagoons can greatly reduce the discharge of ammonia and other gases. Conserving nitrogen in tank and lagoon waste increases the value of the effluent as a fertilizer, but can increase the cost of managing manure to protect water quality.
- ***Solid-liquid separation***—Separating urea from solid fecal matter using sedimentation basins or mechanical methods avoids some of the reactions that cause the formation of ammonia and odor. Separation also reduces the cost of moving waste to land for efficient disposal.
- ***Manure incorporation/injection***—Rapidly incorporating manure into the soil after spreading by plowing or disking—or injecting manure liquids or slurries directly into the soil—reduces odor, ammonia emissions, and the potential for runoff to surface waters. However, incorporation/injection may also increase the risk of nitrogen leaching to ground water.
- ***Comprehensive nutrient management***—Nutrient management matches the combined nutrient applications from manure and commercial nutrient sources to crop needs so that as few nutrients as possible are lost to the environment.

An important characteristic of most of these practices is that in reducing one type of emission, they may increase another type of emission. Such interactions can have an important bearing on the design of policies for protecting environmental quality.

Policy Responses

Federal, State, and local governments have responded to the environmental problems posed by animal operations through a variety of regulations and conservation programs (see Chapter 27, "Federal Laws Protecting Environmental Quality"). The Environmental Protection Agency introduced new Clean Water Act regulations in 2003 for controlling runoff of manure nutrients from the largest animal feeding operations. Concentrated animal feeding operations (CAFOs, defined as those operations requiring a pollution discharge permit) develop and implement a nutrient management plan that bases nutrient applications on agronomic rates. This provision requires CAFOs to spread their manure over a much larger land base than they are currently using, and most will need to move their manure off farm. Livestock and poultry farms'annual net income could decline by more than $1 billion (3.2 percent) if crop producers are reluctant to use manure as a nutrient source (Ribaudo et al., 2003).

USDA is using voluntary approaches such as education and financial incentives to encourage improved manure handling practices on all animal feeding operations (AFOs). Sixty percent of Environmental Quality Incentive Program (see Chapter 24, "Working-Land

Conservation Programs") funds are earmarked to environmental concerns on animal operations.

Many States have enacted regulations that address environmental issues associated with AFOs, including some not addressed at the Federal level. Some States had manure land application requirements in place prior to EPA's 2003 regulations, with coverage often extended to smaller AFOs. Odor is a persistent local issue, and many States are using setback requirements to separate animal operations from residential areas. Ammonia emissions from large animal feeding operations have prompted California to enact regulations in the San Joaquin Valley to protect heavily populated areas downwind.

REFERENCES

[1] Aillery, M., N. Gollehon, R. Johansson, J. Kaplan, N. Key, and M. Ribaudo (2005). *Managing Manure to Improve Air and Water Quality.* ERR-9, U.S. Dept. Agr., Econ. Res. Serv., Sept.

[2] Battye, R., W. Battye, C. Overcash, and S. Fudge (1994). *Development and Selection of Ammonia Emissions Factors.* Report prepared for Office of Research and Development, Environmental Protection Agency, Aug.

[3] Gollehon, N., M. Caswell, M. Ribaudo, R. Kellogg, C. Lander, and D. Letson (2001). *Confined Animal Production and Manure Nutrients.* AIB-771, U.S. Dept. Agr., Econ. Res. Serv., June.

[4] Goolsby, D.A., W.A. Battaglin, G.B. Lawrence, R.S. Artz, B.T. Aulenbach, R.P Hooper, D.R. Keeney, and G.J. Stensland (1999). *Flux and Source of Nutrients in the Mississippi-Atchafalaya River Basin: Topic 3 Report.* Report submitted to White House Office of Science and Technology Policy, Committee on Environment and Natural Resources, Hypoxia Work Group, May.

[5] Jacobson, L.D., R. Moon, J. Bicudo, K. Janni, S. Noll, G. Shurson, J. Zhu, D. Schmidt, C. McGinley, P. Goodrich, R. Nicolai, C. Clanton, K. Davis, L. Brosseau, J. Bruns, C. Pijoan, and T. Blaha (1999). *Generic Environmental Impact Statement on Animal Agriculture; A Summary of the Literature Related to Air Quality and Odor (H).* Report prepared for Minnesota Environmental Quality Board, Minnesota Dept. of Agriculture.

[6] McBride, W.D., and N. Key (2003). *Economics and Structural Relationships in U.S. Hog Production.* AER-818, U.S. Dept. Agr., Econ. Res. Serv., Feb.

[7] Puckett, L.J. (1994). *Nonpoint and Point Sources of Nitrogen in Major Watersheds of the United States.* Water-Resources Investigations Report 94-4001. U.S. Department of the Interior, U.S. Geological Survey.

[8] Ribaudo, M., N. Gollehon, M. Aillery, J. Kaplan, R. Johansson, J. Agapoff, L. Christensen, V. Breneman, and M. Peters (2003). *Manure Management for Water Quality: Costs to Animal Feeding Operations of Applying Manure Nutrients to Land.* AER-824, U.S. Dept. Agr., Econ. Res. Serv., June.

[9] Smith, R.A., R.B. Alexander, G.E. Schwarz, and M.C. Ieradi (2005). "Effects of Structural Changes in US Animal Agriculture on Fecal Coliform Contamination of Streams." Poster presentation, U.S. Dept. Int., U.S. Geological Survey, Reston, VA.

[10] U.S. Environmental Protection Agency (1998). *National Water Quality Inventory: 1996 Report to Congress.* EPA841-R-97-008. Office of Water, Washington, DC, April.

[11] U.S. Geological Survey (1999). *The Quality of Our Nation's Waters: Nutrients and Pesticides*. U.S. Geological Survey Circular 1225.

[12] U.S. Government Accounting Office, (2000). *Key EPA and State Decisions Limited by Inconsistent and Incomplete Data.* GAO/RCED-00-54.

[13] Van Aardenne, J.A., F.J. Dentener, C.G.M. Klijn Goldewijk, J. Lelieveld, and J.G.J. Oliver (2001). "A 1° - 1° resolution dataset of historical anthropogenic trace gas emissions for the period 1890 – 1990," *Global Biogeochemical Cycles* 15(4):909-928.

In: Agricultural Resources and Environmental Indicators ISBN 978-1-60021-467-7
Editor: Keith Wiebe and Noel Gollehon, pp. 157-166

Chapter 17

IRRIGATION WATER MANAGEMENT

Glenn Schaible and Marcel Aillery

Efficient irrigation systems and water management practices can help maintain farm profitability in an era of increasingly limited and more costly water supplies. Improved water management practices may also reduce the impact of irrigated production on offsite water quantity and quality, and conserve water for growing nonagricultural demands. The effectiveness of public water conservation programs depends on how such programs account for diverse farm types and the farm-size characteristics of irrigated agriculture.

WHY MANAGE IRRIGATION WATER?

Agriculture, which accounts for about 90 percent of freshwater consumption in the Western States and over 80 percent nationwide, is increasingly being asked to use less water in order to meet societal demands for other uses (see Chapter 4, "Irrigation Resources and Water Costs"). Water demands are increasing for municipal and industrial uses, recreation, fish and wildlife habitat, and Native American trust responsibilities. For example, conservation of farm irrigation water was a key component of recent water transfer agreements between the Imperial Irrigation District and the San Diego County Water Authority, expected to account for 200,000 acre-feet of annual water transfers during 2021-2047 (Schaible, 2004a).

Farm-level irrigation water management (IWM) involves the managed allocation of water and related inputs in irrigated crop production to enhance economic returns and minimize environmental impacts. USDA identifies improvements in IWM as essential to meeting its national priorities for reducing agriculturally induced nonpoint-source pollution, including surface-and groundwater contamination, reductions in soil erosion and sedimentation, and conservation of ground and surface water (USDA, 2004b). The National Research Council in *A New Era for Irrigation* (NRC, 1996) highlights the importance of IWM "to allocate limited water resources equitably."

Improved IWM can help reduce loadings of nutrients, pesticides, and trace elements in irrigation runoff to surface waters, and leaching of agrichemi-cals into groundwater supplies

(Schaible and Aillery, 2003). Strategies to improve the Nation's water quality (see Chapter 5, "Water Quality: Impacts of Agriculture") must address the effect of irrigation on surface-and groundwater resources (NRC, 1996).

Improvements in IWM can also help maintain the long-term viability of the irrigated agricultural sector. Irrigated cropland is an important and growing component of the U.S. farm economy, accounting for almost half of total crop sales from just 16 percent of the Nation's harvested cropland in 1997

(USDA, 2001). Water savings at the farm level can help offset the effect of rising water costs and limited water supplies on producer income. Improved water management may also reduce expenditures for energy, chemicals, and labor, while enhancing revenues through higher crop yields and improved crop quality. Strategic IWM may also enable producers to better withstand the downside risks of drought.

Use of Improved Irrigation Technology and Management

Producers may respond to limited water supplies through various means, with differing implications for crop production, farm returns, resource use, and environmental quality. Water use per acre may be reduced by applying less than a crop's full consumptive requirement, by shifting to alternative crops or varieties that use less water, or by adopting more efficient irrigation technologies and management practices. Producers may even convert from irrigated to dryland farming or retire land from production.

With water increasingly scarce, irrigators will likely continue to rely on improved technologies and water management practices to conserve water. Irrigation efficiency, broadly defined at the field level, is the ratio of irrigation water beneficially used (crop consumptive use plus an allowance for leaching of salts) to that applied, expressed as a percentage (USDA, 1997).

Irrigation application systems may be grouped under two broad types: gravity flow and pressurized. ***Gravity-flow systems*** distribute water across the field via land treatments—such as soil borders and furrows—that control lateral water movement and channel it in the field. Water is conveyed to the field by means of open ditches, above-ground pipe (including gated pipe and flexible tubing), or underground pipe, and released along the upper end of the field through siphon tubes, ditch gates, pipe valves, or pipe orifices. ***Pressurized systems*** include a variety of sprinkler and low-flow irrigation techniques to distribute water across a field.

With rare exceptions, the pressure to distribute water involves pumping, which requires energy. Sprinkler systems—in which water is sprayed over the field surface, usually from above-ground piping—may be operated on sloping or rolling terrain unsuited to gravity systems. (See the Glossary in the "Irrigation and Water Use" briefing room on the ERS website for more information on these terms.)

Gravity-Flow Irrigation

Although total acreage in gravity systems has declined by 26 percent since 1979, gravity-flow systems still account for 44 percent of irrigated acreage nationwide, down from 63 percent in 1979 (Figure 56). Gravity-flow systems, used in all irrigated areas, are particularly dominant in the Southwest, Central Rockies, Southern Plains, and Delta regions (USDA, 2004a). Furrow application comprises about half of the acreage in gravity-flow systems; border/basin and uncontrolled-flood application account for the remaining acreage (Table 25). Much of the uncontrolled flooding is used for hay and pasture production in the Northern and Central Rockies.

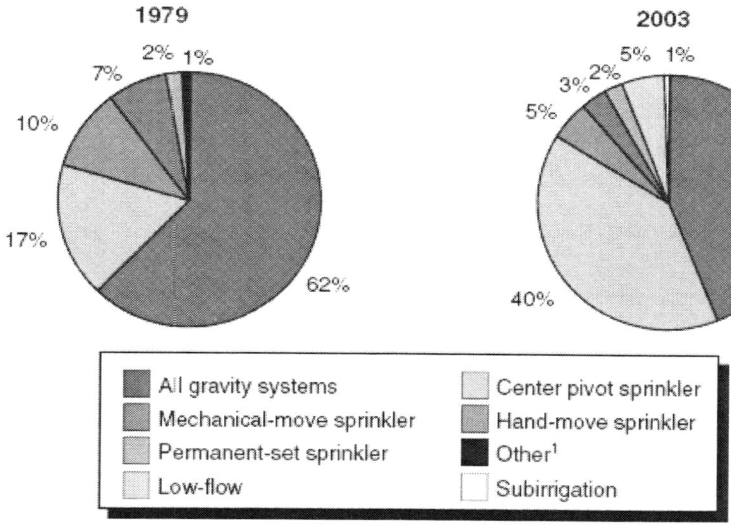

[1]Other for 1979 includes both low-flow and subirrigation.

Source: USDA-ERS, based on Farm and Ranch Irrigation Surveys for 1979 and 2003 (USDC, 1982; USDA, 2004a).

Figure 56. Irrigation systems in 1979 and 2003.

Table 24. Changes in irrigation system acreage, 1979-2003

System	1979	1998	2003	Change 1979-98	Change 1998-2003
		Million acres		Percent[1]	
All systems	50.2	54.2[2]	52.6	8	(3)
Gravity-flow systems	31.2	26.8	23.1	(14)	(14)
Sprinkler systems	18.4	24.6	26.9	34	9
Center pivot	8.6	18.5	21.3	115	15
Mechanical move	5.1	3.0	2.7	(41)	(10)
Hand move	3.7	1.9	1.7	(49)	(11)
Solid/permanent set	1.0	1.2	1.2	20	0
Low-flow irrigation (drip/trickle and micro-spray)	0.3	2.2	3.0	633	36
Subirrigation	0.2	0.6	0.3	200	(50)

[1]Numbers in ( ) indicate a decrease.

[2]Based on USDA-NASS 2004 revised estimate for 1998 due to re-weighting for undercoverage. (The sum of subcategories will differ slightly from aggregates because of rounding error.)

Source: USDA-ERS, based on Farm and Ranch Irrigation Surveys for 1979, 1998, and 2003 (USDC, 1982; USDA, 1999, and USDA, 2004a).

Water losses are comparatively high under traditional gravity-flow systems due to percolation losses below the crop-root zone and to surface-water runoff. Field application efficiencies typically range from 40 to 65 percent, although improved gravity systems with proper water management may achieve efficiencies of up to 80-90 percent (USDA, 1997).

Table 25. Irrigation application systems, by type, 1998 and 2003.

	1998		2003	
	Area	Share of all systems[1]	Area	Share of all systems[1]
	Million acres	Percent	Million acres	Percent
All systems	54.2	100	52.6	100
Gravity-flow systems[2]	26.8	50	23.1	44
Row/furrow application	13.8	25	11.7	22
Open-ditch delivery systems	4.6	9	4.4	9
Pipe/poly-tubing delivery systems	9.2	17	7.4	14
Border/basin application	8.3	15	8.8	17
Open-ditch delivery systems	4.8	9	5.5	10
Pipe/poly-tubing delivery systems	3.5	7	3.3	5
Uncontrolled flooding application	3.2	6	2.3	4
Open-ditch delivery systems	2.8	5	2.1	4
Pipe/poly-tubing delivery systems	0.4	1	0.1	*
Othergravity (mostly with unlined ditches)	1.5	3	0.3	*
Sprinkler systems[2]	24.6	45	26.9	51
Center-pivot	18.5	34	21.3	41
High-pressure (60 psi ormore)	1.9	4	1.9	4
Medium-pressure (30 to 59 psi)	7.4	14	9.7	18
Low-pressure (under 30 psi)	9.2	17	9.7	18
Other sprinkler systems	6.1	12	5.6	9
Low-flow irrigation (drip or trickle)	2.2	4	3.0	6
Subirrigation	0.6	1	0.3	*

[1]Numbers may not add due to multiple systems on some irigated acres and incomplete survey responses.

[2]For a more detailed breakout of irrigation systems, see the ERS Briefing room on "Irrigation Water Management" on the ERS website.

- = less than 1 percent. Source:
- USDA-ERS, based on Farm and Ranch Irrigation Surveys for 1998 and 2003 (USDA, 1999 and 2004a).

Various land treatments, system improvements, and water management measures have been developed to reduce water losses under gravity-flow systems. For example, precision laser-leveled irrigation is practiced on 3.7 million acres (16 percent of gravity acres), mostly in the Southwest, Delta, and Northern Rockies (Montana, Idaho, and Wyoming) regions. Improved gravity systems generally involve on-farm water conveyance upgrades that increase uniformity of applied water and reduce percolation losses and field runoff. However, open-ditch systems still account for 53 percent of gravity acreage served (Table 25; USDA, 2004a). Improved ditch systems, lined with concrete or another impervious substance, account for only 20 percent of gravity acres served by open ditches. Above-ground, pipeline delivery systems—including gated pipe and flexible (poly or lay-flat) tubing—account for 34 percent of all gravity acreage served, with underground pipe delivery systems serving the remaining 13 percent. Surge-flow and cablega-tion systems—designed to control water deliveries from gated pipe—were used on 0.4 million acres, representing 2 percent of gravity-flow acres in 2003 (Figure 57).

Use of improved water management practices for gravity irrigation, while increasing, remains an area of significant growth potential. Alternate-row irrigation is practiced on only 11 percent of gravity-flow acres; special furrowing practices (wide-spaced, compacted, or diked) on 6 percent; and shortened-furrow water runs on 2 percent. Tailwater-reuse pits, designed to recirculate field drainage flows, are used on about 7 percent of gravity acres, while reduced irrigation set-times are observed on 12 percent. Polyacry-lamide—a water-soluble soil amendment designed to reduce soil erosion, enhance water infiltration, and improve nutrient uptake—is used on 2 percent of gravity-flow acres.

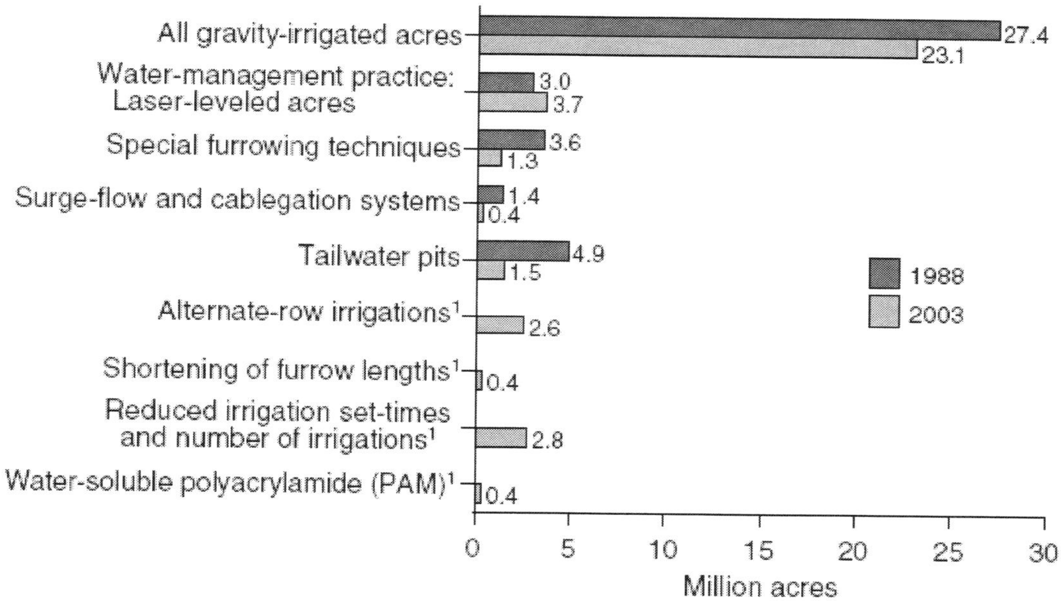

[1]Not reported for 1988.

Source: USDA-ERS, based on Farm and Ranch Irrigation Surveys for 1988 and 2003 (U.S. Dept. Commerce, 1990; USDA, 2004a).

Figure 57. Improved gravity water management practices, 1988 and 2003.

PRESSURIZED IRRIGATION

Sprinkler irrigation has been adopted in many areas as a labor- and water-conserving alternative to gravity-flow systems. Field application efficiencies for properly designed and operated sprinkler systems range from 50 to 95 percent, with most systems achieving 75 to 85 percent (USDA, 1997). Acreage for all pressurized systems expanded from 19 million acres (37 percent of total irrigated acreage) in 1979 to 30 million acres (57 percent) in 2003 (Table 24). Sprinkler systems alone accounted for 27 million acres, or 51 percent of all irrigated acreage in 2003 (Table 25). Acreage in sprinkler systems has continued to expand in recent years, with an increase of nearly 9 million acres (46 percent) since 1988 (USDC, 1990; USDA, 2004a).

Center-pivot sprinkler systems accounted for roughly 79 percent of sprinkler acreage in 2003, or 41 percent of total irrigated acreage (Table 25), increasing by nearly 13 million acres from 1979. Nearly two-thirds of the increase is attributable to net increases in irrigated area under sprinkler, while about a third reflects the net replacement of other sprinkler types with center-pivot systems (Table 24). The more advanced low-pressure center-pivot and linear-move systems, including low-energy precision application (LEPA) systems (below 30 pounds per square inch), combine high application efficiencies with reduced energy and labor requirements. These systems account for 46 percent of center-pivot acreage, and are especially popular in the Southern Plains where irrigation relies heavily on higher-cost ground-water pumping. Current advances in sprinkler technology focus on the variable application of spray heads, as well as remote control of individual sprinklers and nozzles for precision agriculture.

Low-flow systems—including drip, trickle, and micro-sprinklers (with application efficiencies of 95 percent or greater)—were used on 3 million acres in 2003, or just 6 percent of irrigated cropland acreage (Table 25), up from 300,000 acres in 1979 (Table 24). The annual rate of growth (7 percent) was slower during 1998-2003 than the explosive 74-percent rate during 1979-88 (Table 24). Low-flow systems are most commonly used for vegetables and perennial crops such as orchards and vineyards (primarily in California and Florida), although experimentation and limited commercial applications are occurring with some row crops (e.g., cotton).

IRRIGATION SCHEDULING AND WATER-FLOW MEASUREMENT

Proper irrigation scheduling and precise measurement of water flow help producers match water applied to crop needs. Most irrigated farms continue to use a combination of less sophisticated methods to schedule irrigations (USDA, 2004a). Nearly 80 percent of irrigated farms use mere visual observation to evaluate the "condition of the crop," while some farms (ranging from 6 to 35 percent) simply "feel-the-soil," irrigate "when their neighbor irrigates," use a "personal calendar schedule," use "media daily weather/crop evapotranspiration (ET) reports," or irrigate consistent with "scheduled water deliveries." Most irrigated farms do not use the more advanced, information-intensive methods to schedule irrigation; less than 8 percent of irrigated farms use soil and/or plant moisture sensing devices, commercial or government-sponsored irrigation scheduling services, or computer simulation models. These current statistics suggest a significant potential for greater agricultural water conservation through public policy that promotes broader understanding and more extensive application of such scheduling techniques.

Water-flow measurement devices, for both on- and off-farm conveyance, include weirs, flumes, and in-canal flow meters for open ditches, internal/external meters for pipe delivery systems, and flow meters in wells to monitor groundwater pumping. Of the 380,000 wells used in 2003 to pump ground water for agriculture, only 61,000 (16 percent) used flow meters. While this is a 32-percent increase since 1994, flow meters on wells account for just 1 in 5 acres irrigated with ground water.

POTENTIAL FOR IMPROVEMENT IN IRRIGATION CONSERVATION

Significant potential still exists for expanding agricultural water conservation. How much can be achieved depends on the combined use of conserving water-management practices and irrigation systems (Schaible, 2004b; USDA, 2004a). Of the 23.1 million gravity-irrigated acres in 2003, only 56 percent benefited from the use of one or more water management practices—accounting for just 53 percent of gravity-irrigated farms (USDA, 2004a). While not all water management practices can (or should) be applied to all gravity-irrigated acres simultaneously, at least 40-60 percent of gravity irrigation could benefit from improved water management (Figure 57). In addition, while use of low-pressure sprinkler systems increased to 38 percent of total irrigated acres in 2003, at least 39 percent of irrigated acreage likely remains available for improved conservation (Figure 58). The combined effect of improved systems and water management practices, along with more extensive use of advanced irrigation scheduling and water-flow measurement practices across all irrigation, would likely translate to even greater agricultural water conservation potential.

FARM SIZE AND WATER CONSERVATION

An ERS analysis of structural characteristics of Western irrigated farms found that size matters in how well water conservation programs serve both USDA conservation and small-farm policy goals (Schaible, 2004b). In the 17 Western States, which account for 77 percent of U.S. irrigated acres, nearly 81 percent of irrigated farms are small - with less than $250,000 in annual farm sales (FS) (Figure 59). However, large irrigated farms (FS > $250,000) account for 61 percent of irrigated crop acres, nearly 85 percent of irrigated farm sales, and 66 percent of the total farm water applied. The largest 9.5 percent of irrigated farms (FS > $500,000) account for 48 percent of total farm water applied. Average annual water applied ranges from less than 150 acre-feet for the smallest irrigated farms (FS < $100,000) to more than 2,500 acre-feet for the largest farms.

In aggregate, "water-conserving/higher-efficiency" irrigation in the West ranges from 46-78 percent of acreage for pressurized (sprinkler) irrigation to 40-57 percent of acreage for gravity irrigation (Schaible, 2004b). For both categories, relative conservation improvement potential is generally greater for smaller irrigated farms. However, larger farms irrigate many more acres, so aggregate water savings due to a conservation program could be much greater for these farms. While "perceived economic benefits" and "lack of financing ability" are two commonly reported barriers to irrigation system improvements across all irrigated farms, "not investigating the merits of system improvements" is an additional critical barrier to system improvements for smaller irrigated farms.

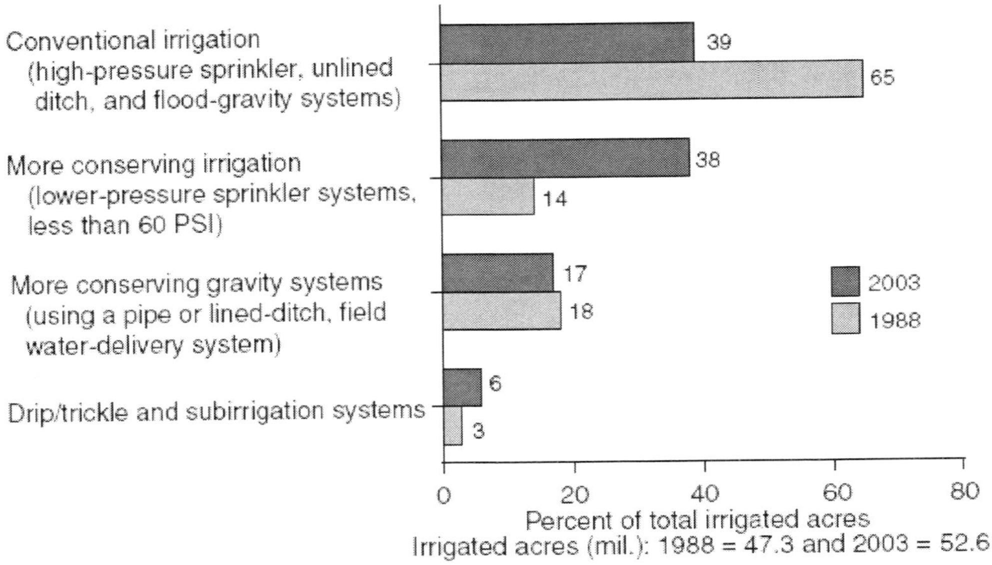

Source: USDA-ERS, based on Farm and Ranch Irrigation Surveys for 1988 and 2003 (U.S. Dept. Commerce, 1990; USDA, 2004a).

Figure 58. Adoption of water-conserving irrigation systems, 1988 and 2003.

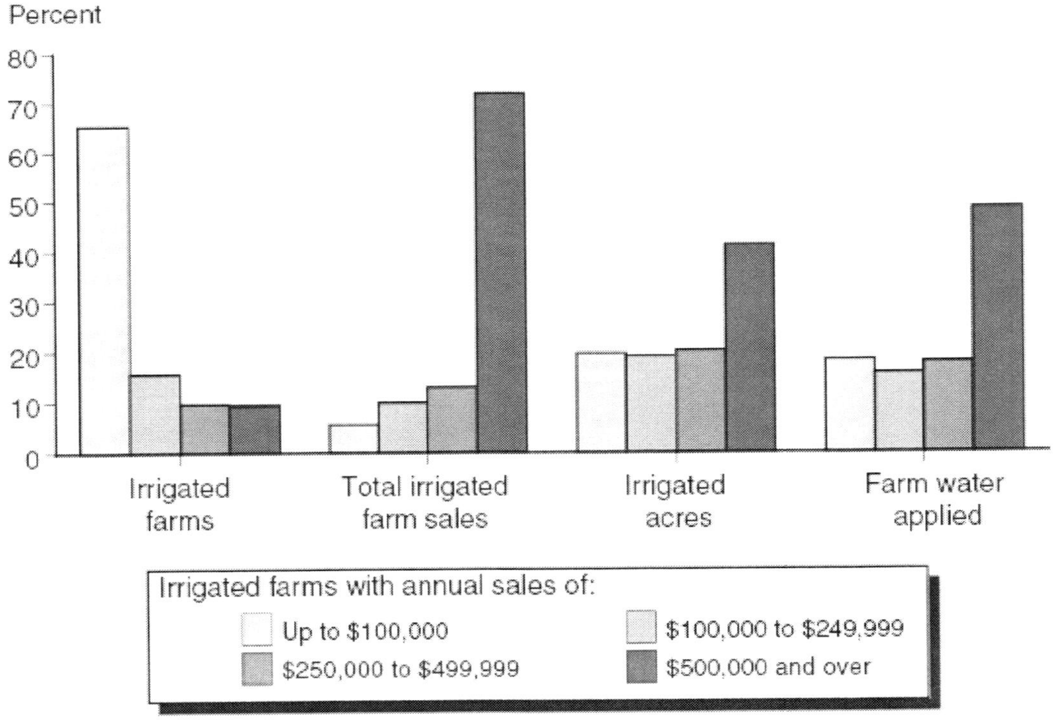

Source: USDA-ERS, based on the Farm and Ranch Irrigation Survey (1998) (USDA, 1999; Schaible, 2004b).

Figure 59. Characteristics of irrigated farms by size class, 17 Western States, 1998.

PRODUCERS' INCENTIVES

While survey results demonstrate that irrigators do implement irrigation system improvements to meet environmental goals, improved farm returns is likely the dominant motivating factor (Table 26). From a private economic perspective, producers generally invest in improved irrigation technologies when perceived benefits are greater than additional (net) producer costs. However, Kim et al. (2000) demonstrate that from a public perspective where water quality benefits accrue largely off-farm, public cost-share funding of a more conserving technology may be warranted. For example, in Merrick County, NE, adoption of tailwater recovery or surge-flow gravity systems may be more profitable to the producer although, even with these systems, groundwater quality would continue to deteriorate. A center-pivot sprinkler system would significantly reduce the accumulation of nitrates in ground water after 15 years. However, adoption of center-pivot systems would reduce producer profits by about $9 per acre (in 1990 dollars), so cost-sharing or other incentives might be necessary to encourage adoption of systems that contribute more to improving water quality.

Table 26. Producer reasons for irrigation conservation improvements, 1999-2003

Irrigated farms implementing irrigation improvements during 1999-2003:	**Farms**
	70,336
	Percent
Reason for/effect of improvements:	
Improved crop yield or quality	57.6
Reduced energy cost	39.0
Reduced water applied	58.5
Reduced labor cost	39.2
Reduced fertilizer/pesticide losses	14.2
Reduced soil erosion	30.8
Reduced tailwater	22.9
Other	8.4

Source: USDA-ERS, based on the Farm and Ranch Irrigation Survey (2003), Vol. 3, Special Studies, Part I, table 39 (USDA, 2004a).

Federal, State, and local cost-share programs that address farm water delivery, field-level irrigation systems, and farm water management practices are key to improving irrigation efficiency. Only about 13 percent of irrigated farms in the West participated in public cost-share programs for water conservation between 1994 and 1998. Smaller irrigated farms make up 77 percent of participants in USDA cost-share programs designed to encourage irrigation or drainage improvements. Given that such farms account for only 34 percent of farm water applied in the West, these results indicate that farm size matters in the effectiveness of current agricultural water-conservation programs. Cost-share programs that target larger farms would likely conserve more water, making more water available to meet environmental and other objectives, especially when integrated with State-sponsored water markets, water banks, and conserved-water-rights programs (Schaible, 2004b). Integrated Federal/State conservation policy would likely increase opportunities to better balance alternative farm policy objectives—i.e., resource efficiency and potential gains in water saved, with distributional considerations involving cost-share funding allocations.

REFERENCES

[1] Kim, C.S., G.D. Schaible, and S.G. Daberkow (2000). "An Efficient Cost-Sharing Program to Reduce Nonpoint-Source Contamination: Theory and an Application to Groundwater Contamination," *Environmental Geology*, Vol. 39, No. 6 (April): pp. 649-659.

[2] National Research Council (NRC) (1996). *A New Era For Irrigation*, Water Science and Technology Board—Committee on the Future of Irrigation in the Face of Competing Demands, National Academy Press, Washington, DC.

[3] Schaible, Glenn D., and Marcel P. Aillery (2003). "Irrigation Technology Transitions in the Mid-Plains States: Implications for Water Conservation/Water Quality Goals and Institutional Changes," *International Journal of Water Resources Development*, Vol. 19, No. 1, pp. 67-88.

[4] Schaible, Glenn D. (2004a). "Agricultural Risks in a Water-Short World: Producer Adaptation and Policy Directions—A Workshop Summary," www.farmfoundation.org/documents/Z4C1-WaterWorkshopSummary-Final-V1c_11-8-04.pdf

[5] Schaible, Glenn D. (2004b). "Western Irrigated Agriculture: Farm-Size Characteristics." Data product on the ERS website, U.S. Dept. Agr., Econ. Res. Serv.

[6] U.S. Department of Agriculture (1997). *National Engineering Handbook*, Part 652, Irrigation Guide, Natural Resources Conservation Service.

[7] U.S. Department of Agriculture (1999). *Farm and Ranch Irrigation Survey (1998), Vol. 3, Special Studies-Part 1 of the 1997 Census of Agriculture*, AC97-SP-1, Natural Agricultural Statistics Service.

[8] U.S. Department of Agriculture (2001). "Irrigation and Water Use" Briefing Room on the ERS website, Econ. Res. Serv.

[9] U.S. Department of Agriculture (2004a). *Farm and Ranch Irrigation Survey (2003), Vol. 3, Special Studies-Part 1 of the 2002 Census of Agriculture*, AC-02-SS-1, National Agricultural Statistics Service, (Nov.).

[10] U.S. Department of Agriculture (2004b). *Farm Bill 2002: Environmental Quality Incentives Program*, Fact Sheet, Natural Resources Conservation Service (Oct.). (See also Public Law 107–171–May 13, 2002, *Farm Security and Rural Investment Act of 2002*, Subtitle D– Environmental Quality Incentives, Sections 2301 and 1240I.)

[11] U.S. Department of Commerce (1982). *1979 Farm and Ranch Irrigation Survey*, Vol. 5, Special Reports—Part 8, AC78-SR-8, Bureau of the Census.

[12] U.S. Department of Commerce (1990). Farm and Ranch Irrigation Survey (1988), Vol. 3, Part 1—Related Surveys of the 1987 Census of Agriculture, AC87-RS-1, Bureau of the Census.

In: Agricultural Resources and Environmental Indicators ISBN 978-1-60021-467-7
Editor: Keith Wiebe and Noel Gollehon, pp. 167-173

Chapter 18

INFORMATION TECHNOLOGY MANAGEMENT

Stan Daberkow, Mitchell Morehart, and William McBride

Information technology (IT) is affecting the way farmers produce and market their output and how rural residents receive services and communicate. While computers and the Internet are the most common IT tools in use today, IT also encompasses software and associated services, such as telecommunications, required to fully use these technologies.

INTRODUCTION

Information technology (IT) enables U.S. farmers to access real-time market information and buy and sell through e-commerce sites; manage their cropland at ever smaller scales (to meet both economic and environmental objectives) through precision agriculture; and use modern accounting, recordkeeping, and tax management through computer and Internet resources.[1] Telecommunication infrastructure in rural areas is crucial if farmers and rural residents are to adopt and utilize IT. Many government agencies, including those servicing farmers, are offering clients the ability to receive information and program benefits via the Internet.

INFORMATION TECHNOLOGIES FOR FARM MANAGEMENT DECISIONS

IT adoption by U.S. farms has exhibited significant growth over the last several years; as of 2003, about half of all farms had computer and/or Internet access (Figure 60). However, only about 30 percent of the farms reported using a computer for the farm business. Internet access grew from less than 15 percent of all farms in 1997 to 48 percent in 2003, and 5 percent of all producers reported using the Internet to contact a USDA website.

[1] Information technology is broadly defined as those technologies that allow individuals to create, seek, and manipulate information (Vanderheiden and Zimmermann, 2002).

Farm IT Users and Uses

Periodically, information on computer and Internet use is collected in the Agricultural Resource Management Survey (ARMS). The 1999 ARMS measured the extent of farmers'Internet use and online purchases/sales of farm products. Many agricultural e-commerce ventures were just getting started in 1999, so this was a first look at how farm businesses were using IT. Farms that bought or sold online in 1999 were more likely to be run by younger, more educated operators than the national average. Almost three-quarters of active e-commerce users were between 35 and 54 years old, and just over a third had completed college or graduate school. Higher rates of adoption among these groups are to be expected, since the willingness to adopt new technologies is often related to both age and education. Over 42 percent of farmers'online market activity in 1999 involved purchasing crop inputs (e.g., seed, fertilizers, and pesticides), and online buying was related to farm size. In contrast, farm size showed no relation to online purchasing of livestock inputs (e.g., feed and feeders) and selling of livestock (58 percent of farmers'online market activity).

Source: USDA, National Agricultural Statistics Service
http://usda.mannlib.cornell.edu/reports/nassr/other/computer/fmpc0703.pdf

Figure 60. U.S. farms using computers, 1997-2003.

The 2000 ARMS was extended to examine the types of activities that were conducted online. During 2000, producers reported \$665 million in online buying and selling. Online purchases totaled \$378 million, covering machinery and equipment, farm supplies, crop inputs, livestock inputs, and office and computer equipment. Purchases of crop and livestock inputs together were 35 percent of total online purchases, and each was smaller than machinery and equipment purchases and general farm supply purchases. Online sales by farmers totaled \$287 million—\$191 million in livestock sales and \$96 million in crop sales.

Farmers reported using the Internet for various management activities. The most common use was price tracking, reported by 82 percent of Internet users. Information gathering from government and other sources was also relatively common. Communication with other farmers and advisory services was reported by about 30 percent of Internet users. The least often reported Internet activity was the management of business finances such as online banking, paying bills, and obtaining loans.

In 2002, ARMS investigated the intensity of business/personal use of the Internet by U.S. farmers. (Internet use was conditioned on the operator's reporting computer use.) Internet use was positively related with farm size. The share of farms using the Internet in their business ranged from 16 percent of limited-resource farms to nearly 75 percent of very large farms (Figure 61). Time spent on the Internet for farm business purposes also increased with farm size. Only 20 percent of operators over age 65 reported Internet use, versus over half of operators between age 35 and 44. Farms that specialized in crops were more likely to use the Internet than were livestock operations. Half of the farmers reporting Internet use reported that they spent 6 hours or less per week online. Fewer than 10 percent of Internet users spent 20 hours or more per week online.

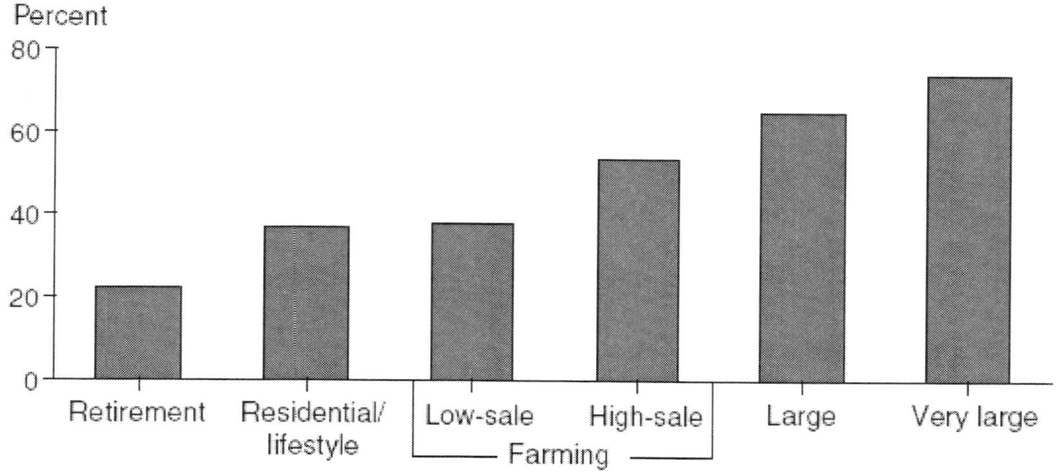

Source: USDA, ERS, (http://www.ers.usda.gov/Briefing/ARMS/)

Figure 61. Internet use for farm business by farm type, 2002.

INFORMATION TECHNOLOGIES FOR CROP PRODUCTION

Recent advances in the computer, aerospace, and communications industries allow farmers to monitor and manage soils and crops on small areas of individual fields. Precision agriculture or site-specific crop management are the terms often applied to the suite of information technologies used for sensing subfield spatial and temporal variability and customizing applications across the field. A number of spatially oriented information technologies are commercially available for most crops to help with fertilizer, pesticide, seed, irrigation, and tillage decisions. Rather than treat fields uniformly, producers can use these technologies to manage soil, pest, landscape, or microclimate variability by adjusting input use within a field to enhance returns and to reduce potential environmental risks. Such technologies include yield monitors; the Global Positioning System (GPS); Geographic Information Systems (GIS); guidance systems; satellite, aerial, and on-the-go sensors; and variable-rate applicators

Adoption Trends

Based on annual USDA-ARMS surveys of corn, soybean, wheat, and cotton producers, the adoption of precision agriculture (PA) technologies varied widely across these major crops between 1996 and 2003 (Table 27). Yield monitors are the most widely used PA technology, reaching over 35 percent of all corn acres (2001) and nearly 30 percent of all soybean acres (2002). This technology became commercially available to grain producers in the early 1990s, but did not become available to cotton growers until the late 1990s. Only about a third of the corn and soybean acres on which yield monitors were used were connected to the GPS and generated a yield map—an indication that producers have been cautious about using this technology for changing production practices.

Table 27. Share of U.S. corn, soybean, wheat, and cotton acres on which yield monitors and yield maps were used, 1996-2003[1]

Technology/year	Corn	Soybeans	Wheat	Cotton
	Percent of planted acres			
Yield monitor				
1996	15.6	13.3	5.9	NA
2000	34.2	25.4	9.1	1.3
2001	36.5	NA	NA	NA
2002	NA	28.7	NA	NA
2003	NA	NA	NA	2.6
Yield map				
1996	NA	8.1	*	NA
2000	13.8	7.8	*	*
2001	13.7	NA	NA	NA
2002	NA	10.7	NA	NA
2003	NA	NA	NA	1.7

NA = survey not conducted. * = less than 1 percent.

[1]These estimates are revised from previous published estimates based on updated weights from the ARMS.

Source: For more information, see ARMS Briefing Room on the ERS website.

Remote sensing, variable-rate applicators, and guidance systems are among the most recent, as well as most rapidly evolving, precision agriculture technologies. Geo-referenced soil data, such as pH or nitrate levels and soil type, can also help producers intensely manage their crops. Recent ARMS data indicate that the adoption of these technologies, like yield monitors and mapping, differs by crop. Remote sensing, either by airplane or satellite, was reportedly used on less than 10 percent of planted acreage in recent years. While remote sensing can detect variation in vegetative reflection, the cause of that variation may still require confirmation on the ground. Also, cost, timeliness, and image resolution issues may be inhibiting the spread of this technology.

Machine guidance systems, which are connected to GPS, were introduced in the late 1990s and producers reported using these systems on 6-7 percent of corn and soybean acres during 2001-02, and on over 10 percent of cotton, barley and sorghum acres during 2003. Such systems can reduce costs associated with equipment skips and overlap; permit operation in dust, fog, and darkness; help manage soil compaction; and reduce driver fatigue. Variable-rate technologies (VRT) allow the application of inputs at different rates based on agronomic (or economic) factors that vary within a field. Variable rate application of fertilizer on corn and soybeans was the most widely reported use of this technology (Table 28). Producers reported using VRT to apply inputs on less than 5 percent of planted wheat and cotton acres.

Table 28. Share of U.S. corn, soybean, wheat, and cotton acres on which variable rate technologies were used to apply major inputs, 1998-2003

Year	Corn			Soybeans		
	Fertilizer	Seed	Pesticides	Fertilizer	Seed	Pesticides
	Percent of planted acres					
1998	12.3	4.1	2.4	6.7	*	*
1999	17.5	4.2	1.1	8.3	2.0	1.7
2000	14.5	4.5	3.8	5.8	2.5	1.0
2001	9.8	2.4	3.8	NA	NA	NA
2002	NA	NA	NA	5.0	*	1.3
	Wheat			Cotton		
1998	2.6	1.5	1.7	2.0	1.3	1.5
1999	NA	NA	NA	1.0	1.8	2.0
2000	3.1	*	*	3.8	2.4	2.7
2003	NA	NA	NA	3.9	*	1.9

* = less than 1 percent. NA = survey not conducted.

Source: For more information, see ARMS Briefing room on the ERS website.

Table 29. Share of U.S. acreage on which precision agriculture technology was used, select crops and years[1]

Technology	Sunflower 1999	Potatoes 1999	Sugarbeets 2000	Rice 2000	Barley 2003[2,3]	Sorghum 2003[2,3]
Yield monitor	17.1	10.4	1.0	17.6	17.0	14.4
Yield map	3.8	10.2	*	5.1	4.6	2.0
Geo-referenced soil map	3.8	18.7	28.6	9.5	7.3	7.3
Remote sensing	4.4	20.5	35.2	4.7	2.8	4.4
VRT used for:						
Fertilizer/lime	2.8	13.1	11.9	1.6	12.9	4.7
Seed	*	1.5	2.2	1.2	8.0	3.5
Pesticides	*	3.6	1.3	2.6	10.4	2.7
Guidance	NA	NA	NA	NA	14.7	10.4

* = less than 1 percent. NA = survey not conducted. VRT = variable-rate technology.

[1]These estimates are revised from previous published estimates based on updated weights from the ARMS.

[2]Prior to 2002, respondents were asked if the soil characteristics of the field had ever been geo-referenced. Beginning in 2002, respondents were asked about geo-referencing in the current and previous year.

[3]The question was reworded in 2002 to better define the term "remotely sensed."

Source: For more information, see ARMS Briefing room on the ERS website.

Producers of high-value crops (i.e., sugarbeets and potatoes) tend to use precision agriculture—particularly variable rate fertilizer application—on a higher share of crop acreage than field crop producers (Table 29). Sugar-beet producers, especially in the Red River Valley, reported relatively high use of geo-referenced soil maps and remote sensing in 2000; this is related to the importance of nitrogen management in sugarbeet profitability (Daberkow et al., 2003).

Factors Influencing Adoption

A number of factors—such as profitability, farm and farm operator characteristics, university research and extension activities, and government agency use of IT—will likely affect adoption trends in precision agriculture (PA). Most studies of PA technologies have shown positive economic benefits from the adoption. For example, Lambert and Lowenberg-DeBoer (2000) reviewed 108 PA studies and found 63 percent of the studies indicated positive net returns for a given PA technology, 11 percent reported negative returns, and 26 percent indicated mixed results. Much of the current research indicates that larger farms, located in the Corn Belt and operated by producers familiar with computers, have a higher probability of adopting precision agriculture technologies than farms without such characteristics (Daberkow and McBride, 2000).

Numerous land-grant universities have established PA research and extension programs geared toward adapting IT for crop and livestock production and reducing agriculture's impact on the environment. Universities in Arizona, Mississippi, and Utah are participating in NASA's Space Grant Extension Specialist in Geospatial Technology pilot program to explore how to meet the needs of farmers, ranchers, planners, and others involved in agriculture, natural resource management, and rural development. Similar in scope is the Upper Midwest Aerospace Consortium (UMAC), consisting of participants from North and South Dakota, Montana, Wyoming, and Idaho. USDA's Natural Resources Conservation Service and Farm Service Agency are beginning to offer geo-referenced, field-level data specifying soil types and field boundaries, some of which can be accessed over the Internet. Many farmers can also obtain commodity and conservation program information via the Internet.

Federal IT Policies for Agriculture and Rural Areas

Several Federal policies may facilitate the development and adoption of PA technologies and IT-related services. For example, the Conservation Security Program is a voluntary program that provides incentive payments to farmers to implement or maintain conservation practices on working lands (see Chapter 24, "Working-Land Conservation Programs"). Such practices include the use of yield monitors, a stewardship practice that addresses water quality concerns (*Federal Register*, 2005). As communication and information service becomes increasingly important, rural or farm communities lacking such services may be economically disadvantaged. Federal programs addressing these issues are discussed in the "Rural Telecommunication" briefing room on the ERS website.

REFERENCES

[1] Daberkow, S., and W. McBride (2000). "Adoption of Precision Agriculture Technologies by U.S. Farmers," 5th International Conference on Precision Agriculture, Minneapolis, MN.

[2] Daberkow, S., and W. McBride (2003). "Farm and Operator Characteristics Affecting the Awareness and Adoption of Precision Agriculture Technologies in the U.S.," *Precision Agriculture* 4:163-177.

[3] Daberkow, S., W. McBride, and M. Ali (2003). "Implications of Remote Sensing Imagery and Crop Rotation for Nitrogen Management in Sugar Beet Production" selected paper presented at AAEA meetings, July, Montreal, Canada.

[4] *Federal Register* (2005). March 25 (Vol. 70, No. 57)[Notices]—Pages 15277-15283.

[5] Griffin, T.W., J. Lowenberg-DeBoer, D.M. Lambert, J. Peone, T. Payne, and S. Daberkow (2004). "Precision Farming: Adoption, Profitability, and Making Better Use of Data," Purdue University Staff Paper #04-06 and paper presented at the 2004 North Central Region Triennial Farm Management Conference. Lexington, KY, June.

[6] Hopkins, J., and M. Morehart (2001). "Farms, the Internet, & E-Commerce: Adoption & Implications." *Agricultural Outlook* (AO-286), Nov.

[7] Lambert, D., and J. Lowenberg-DeBoer (2000). "Precision Agriculture Profitability Review," Site-Specific Management Center, School of Agriculture, Purdue University, West Lafeyette, IN.

[8] McBride, W., and S. Daberkow (2003). "Information and the Adoption of Precision Farming Technologies," *Journal of Agribusiness* 21, 1 (Spring):21-38.

[9] U.S. Department of Agriculture, National Agricultural Statistics Service (various years) *Computer and Internet use by farmers.*

[10] Vanderheiden, G., and G. Zimmermann (2002). "State of the Science: Access to Information Technologies," inn J.M. Winters, C. Robinson, R. Simpson, and G. Vanderheiden (eds.), *Emerging and Accessible Telecommunications, Information and Healthcare Technologies* (pp. 152-184). Arlington, VA: RESNA Press.

In: Agricultural Resources and Environmental Indicators ISBN 978-1-60021-467-7
Editor: Keith Wiebe and Noel Gollehon, pp. 175-182

Chapter 19

PRODUCTION SYSTEMS MANAGEMENT AND CONSERVATION PRACTICES

William Quinby, Linda Foreman, Janet Livezey and C.S.Kim

Farmers manage soils, pests, nutrients, and other inputs as part of a system of inter-related production and conservation practices, whether in conventional systems (used by most U.S. farmers) or organic systems. Among all U.S. farmers, those who adopt selected conservation practices (such as crop rotation, conservation tillage, scouting for pests, and soil testing) are more likely than non-adopters to be younger, full-time operators who plant more acreage and participate in government programs. (Characteristics of organic farming systems are examined in chapter 4)

PRODUCTION MANAGEMENT SYSTEMS

Production system choices may be motivated by a desire to increase profits, respond to social objectives, or maintain a way of life for future generations. These potentially competing goals are reflected in the choices and amounts of inputs used for production. Agricultural production management deals with how farmers combine land, water, machinery, structures, commercial inputs, labor, and management skills to produce crop and livestock commodities. Management systems embody some of the more important decisions related to production, and include nutrient management, soil management, water management, weed management, and the like. The overall production management system can be thought of as the combination of activities chosen for each aspect of production.

MANAGEMENT SYSTEMS FOR MAJOR FIELD CROPS

Production management for major field crops can be divided into different stages and/or technology suites, among them:

- Soil management systems (see Part 4).
- Rotation—Deciding what crops and varieties to grow, in what sequence, and whether to double-crop, fallow, or plant a cover crop in order to best use the soil's productive capacity.
- Tillage—Deciding how best to prepare the soil for planting while preserving soil, moisture, and nutrients.
- Conservation structures—Deciding what investments in soil conservation structures to undertake to preserve soil, soil moisture, and soil nutrients.
- Pest management systems (see Part 4).
- Weed management—Deciding what resources to use in determining and controlling weed problems; how to combine scouting, tillage, rotation, cultivation, herbicides, and seed variety choices.
- Insect management—Deciding how to determine and control insect problems.
- Disease management—Deciding how to determine and control disease problems.
- Nutrient management systems—Determining soil nutrient needs for crop growth, and the method/timing of applying animal manure, compost, or commercial fertilizer (see Part 4).
- Manure management systems—Determining the manner of collection, containment, field spreading, and other means of manure disposal (see Part 4).
- Water management systems—Determining the water needed for crop growth and the means of enhancing soil moisture to meet those requirements (see Part 4).
- Irrigation—Deciding the technology and management practices that affect water use efficiency, fuel type, source of water, and scheduling of applications.
- Farm management systems—Determining who decides what (see Part 4).
- Information systems—Determining how much to invest inhouse in computer/internet and/or use of various outside sources/professional consultants to improve the effectiveness of management and crop production (see Part 4).
- Precision agriculture—Deciding what human skills and technologies to employ in adjusting inputs as crop needs vary within each field.
- Variable-rate technology—Deciding what technologies to use in automatic adjustment of input use without real-time control by the machinery operator.

The choices within different management areas are not mutually exclusive. A practice decision may include more than one management system. For example, a crop rotation may be an important component of water management, soil management, pest management, and nutrient management systems.

Adoption of Recommended Conservation Practices

Farmers'production choices may be motivated by both private and public goods, including increased profits and protecting the environment. If operators are to manage their production activities to include social objectives, State and Federal communications about recommended conservation practices are critical. U.S. farmers increasingly face both economic and social pressures to adapt management practices to meet conservation goals. For example, the 2002 Farm Security and Rural Investment Act expanded the eligibility and choices for farmers to receive incentive payments for using environmentally sound practices. The Conservation Security Program (CSP), established in the 2002 Act, rewards environmental stewardship practices in nutrient, pest, soil, and water management (see Chapter 24, "Working-Land Conservation Practices").

Farms that adopt more of the recommended practices under CSP or the Environmental Quality Improvement Program (EQIP) differ from less intensive adopters, and achieve different economic/environmental results. The Agricultural Resource Management Survey (ARMS) includes several questions on the adoption of recommended conservation practices. Farmers were grouped by their combined score on representative practices in five aspects of production management (see box, "Index of Recommended Practices"). ARMS data for 1998 wheat, 2001 corn, 2002 soybeans, and 2003 cotton were used to compare high and low adopters of recommended practices on the fields used to produce these crops.

Adoption ranges from only 3 percent of wheat acreage using variable-rate technology to 92 percent of cotton acreage being scouted for pests (Table 30). (This is primarily a reflection of differences in both economic returns from these practices and in conservation needs, and should not be interpreted as an indicator of differences in conservation effort or commitment.) The number of recommended practices used per acre ranges from an average of 1.8 for cotton to 2.4 for soybeans. There is a strong economic incentive to rotate crops for soybeans (84 percent rotated) and corn (80 percent). Farmers who rotate wheat crops tend to fallow their fields for a year in dryer regions, and double-crop, observing a corn-wheat-soy rotation, in warmer regions. Scouting for pests was the most common recommended practice used for wheat (83 percent) and cotton (92 percent). Pest control accounts for a larger proportion of cotton production costs compared with other crops, and scouting helps minimize pest control costs.

Table 30. Percent of acreage with recommended practice, by crop

Practice	Corn	Soybeans	Wheat	Cotton
	Percent of crop acreage			
Crop rotation	80	84	57	27
Conservation tillage	43	69	33	11
Scouted for pests	55	58	83	92
Soil test for nitrogen	26	24	30	37
Variable-rate tech for inputs	11	6	3	15
Avg. number of practices per acre	2.2	2.4	2.1	1.8

Source: USDA's Agricultural Resource Management Survey: 2001 for corn, 2002 for soybeans, 1998 for wheat, and 2003 for cotton.

The number of recommended conservation practices used ranges from an average of 1.3 practices on cotton in the Prairie Gateway to 2.7 practices on wheat in the Southern Seaboard. For each of the four crops, more than 80 percent of the acreage received one to three of the five recommended practices and less than 6 percent received none (Figure 62).

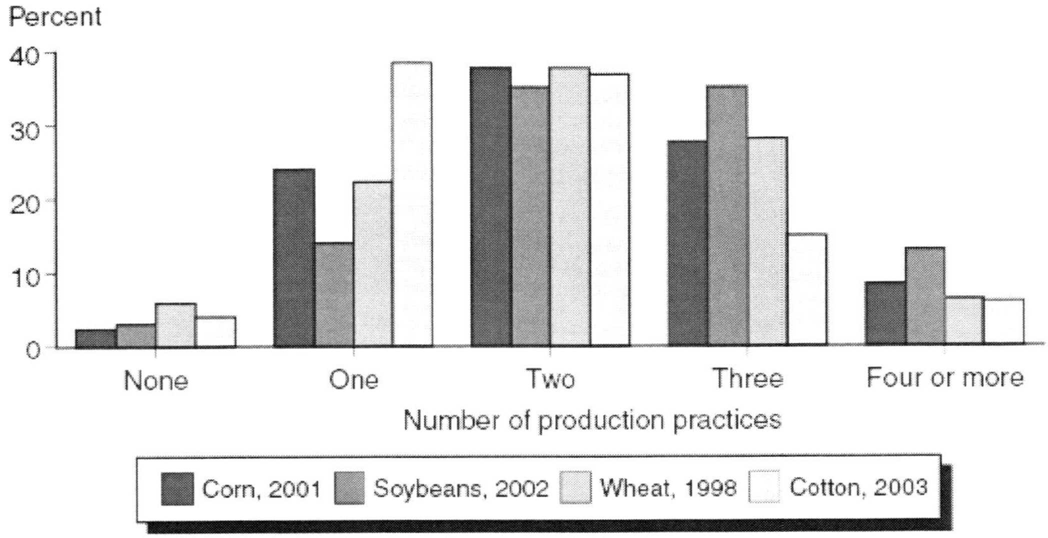

Source: USDA's Agricultural Resource Management Survey.

Figure 62. Distribution of planted acres across number of recommended practices adopted, by crop.

ROLE OF GOVERNMENT PROGRAMS

Corn and soybean producers who participate in government agricultural programs adopt more of the recommended production practices than producers who do not participate. In 2001, corn producers who received program payments used, on average, almost twice as many of the recommended practices as producers not receiving payments. Conversely, operators who adopted one or more of the practices were much more likely (82 percent) to receive government payments than nonadopters (57 percent).

Factors other than program participation influence adoption of recommended practices. Large farms adopt more recommended practices (and are also more likely to participate in programs). Also, any producer with crop-land that contains a wetland or is highly erodible, as defined by the Natural Resources Conservation Service, must use an approved conservation system on that land to receive government payments (see Chapter 23, "Compliance Provisions for Soil and Water Conservation"). They may also benefit from adopting recommended practices, regardless of program requirements, through reduced costs. ARMS data show that farms with wetland or highly erodible land (HEL) adopt more of the recommended practices, and are also more likely to participate in programs. The increased likelihood of wetland or HEL among program participants could explain part of the higher adoption rates for program participants.

Other factors that affect both adoption of approved practices and participation in programs include livestock production, age, education, primary occupation, off-farm occupation, and business structure (see Lambert et al., 2006). Each could explain a part of the higher adoption rates for program participants.

Index of Recommended Practices

For each crop, an index was constructed based on the following practices recommended as contributing to conservation objectives. The total score could range from 0 (adopted none of the practices) to 5 (adopted all of the practices), where 1= used recommended production practice, and 0= did not use recommended production practice.

- **Rotation**—Zero indicates the same crop was planted for 2 consecutive years. All other rotation schemes are scored one. Under this definition, idling or fallowing land during the previous spring and summer counts as rotation. Double cropping is not counted as a rotation if the current crop is the same as the crop planted 1 year prior.
- **Tillage**—One indicates producer used conservation tillage (30 percent or greater residue remaining). Conservation tillage includes no-till, mulch-till, and ridge-till systems.
- **Scouting for pests**—One indicates producer scouted crop for any pests, including weeds, insects, or disease. Casual scouting while in the field for other purposes is counted.
- **Testing for nutrient requirements**—One indicates a soil test for nitrogen or phosphorus was performed, or that a plant tissue test was performed.
- **Use of variable-rate technology**—One indicates that a variable-rate technology was used for applying fertilizer, lime, seeds, or pesticides. Yield, soil, or pest mapping without use of a variable-rate technology is not counted.

FARM AND OPERATOR CHARACTERISTICS

Farms that plant more acreage also use more recommended practices than farms that do not. Farms that use four or five practices typically plant about four times as much corn and about twice as much wheat or soybeans as farms that use none of the practices (Figure 63).

Producers who used more conservation practices were typically younger (Figure 64). Whereas about a third of producers using none of the practices were younger than 50 years old, half of producers that used four or more conservation practices were under 50. Younger producers have longer time horizons for receiving the benefits from conservation practices and are more likely than older producers to make an investment for a long-term payoff. No-till and variable-rate technologies, for example, require large capital investments. Younger producers also have more of an incentive to rotate their crops to keep their field productive since they are more likely to be using the field for many years.

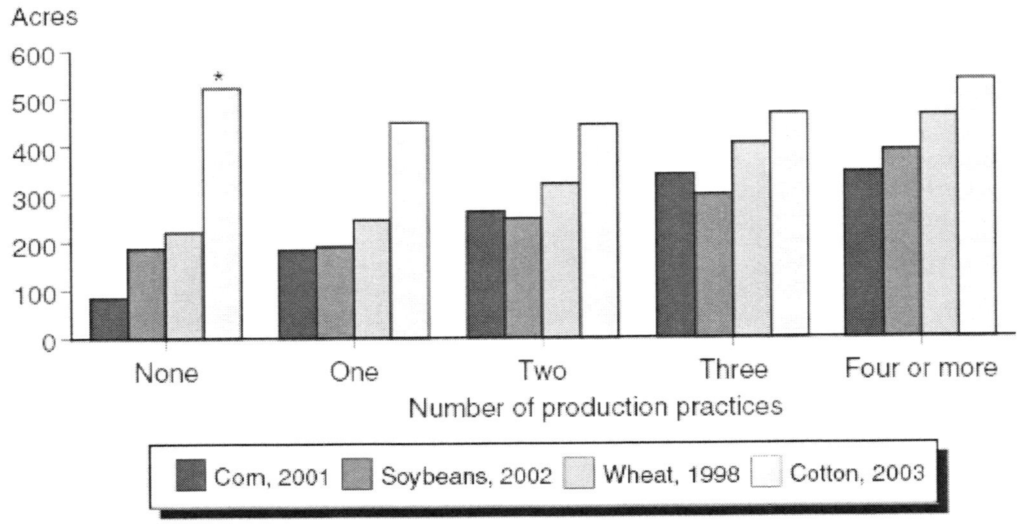

*Small sample size.
Source: USDA's Agricultural Resource Management Survey.

Figure 63. Average acres planted per farm by crop and by number of practices adopted.

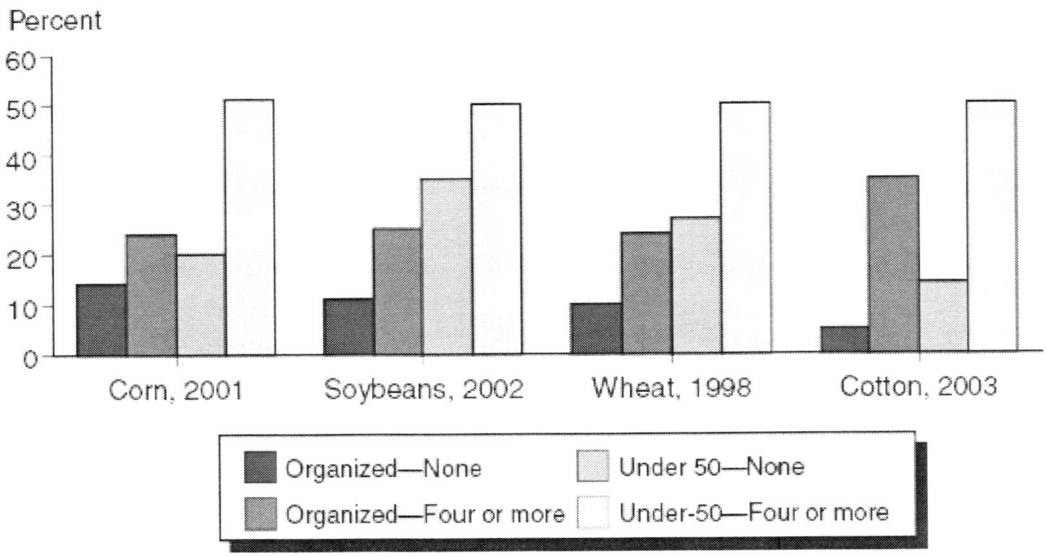

Source: USDA's Agricultural Resource Management Survey.

Figure 64. Producers using four or more recommended conservation practices are more likely to be under 50 and organized as a partnership or corporation.

Producers using more conservation practices are also more likely to operate farms as partnerships or as family corporations rather than sole ownerships (Figure 64). Full-time operators of larger, more complex enterprises may be more likely to have the necessary skills to optimize implementation of newer conservation practices. They also can spread the costs of obtaining information over a larger operation. Producers in partnerships and family corporations may have multiple managers to split the farm management workload, allowing greater depth of knowledge and experience about farm practices. Partnerships and corporations are also more likely to have management successors, giving them a longer time horizon.

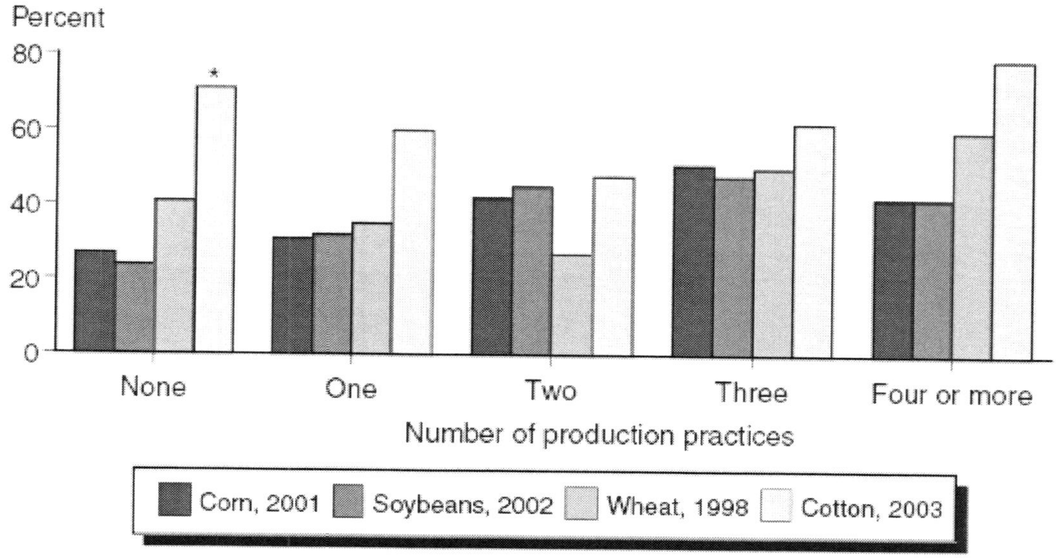

*Small sample size.
Source: USDA's Agricultural Resource Management Survey.

Figure 65. Percent receiving some college education, by crop and by number of recommended conservation practices adopted.

Producers who adopt more recommended practices are more educated, on average (Figure 65). A higher percentage of corn, soybean, and wheat producers who adopted four or more conservation practices completed some college, compared with producers who adopted none. Increased schooling may help producers handle complex farming operations by improving the operator's ability to assimilate new information. Education may also help a producer understand and adapt to changing technologies and recommendations.

Producers who adopt more practices are more likely to be full-time farmers listing farming as their principal occupation. Full-time producers are less likely to have non-farm jobs that compete for their time or provide alternate sources of income. Producers more dependent on farming for income are likely more motivated to explore every possibility to reduce the risk of crop failure or yield reductions. Hence, full-time producers may be more likely to scout their fields for pests, conduct nutrient tests, and stay abreast of the long-term benefits of using conservation practices.

INDICATORS OF CONSERVATION PERFORMANCE

According to ARMS, farmers who adopt more recommended practices generally perform better on conservation objectives. One such objective is to minimize spillover loss of nutrients into the environment. In practical terms, that means reducing the application of nutrients to just what is needed by the crop. A higher ratio of nutrient applied per bushel of grain or bale of cotton lint indicates a higher potential for nutrient contamination of surface and ground waters. ARMS data show that farms using more recommended practices generally apply less total nutrients per unit of product. This is especially true for wheat (Figure 66). High adopters also apply less phosphate on soybeans and less potash on corn. Using fewer inputs both conserves resources and lessens the potential environmental impact from the manufacture, transport, and use of the input.

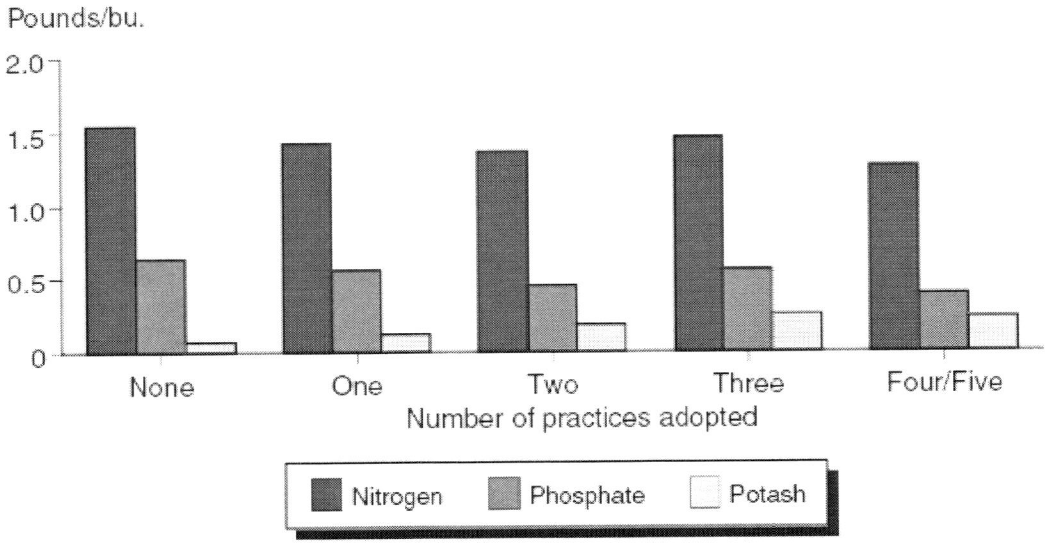

Source: USDA's Agricultural Resource Management Survey.

Figure 66. Nutrient pounds applied per bushel of wheat, by number of recommended practices.

REFERENCES

[1] Lambert, Dayton, Patrick Sullivan, Roger Claassen, and Linda Foreman (2006), *Conservation-Compatible Practices and Programs: Who Participates?* ERR-14, U.S. Dept. Agr., Econ. Res. Serv., Feb.

In: Agricultural Resources and Environmental Indicators ISBN 978-1-60021-467-7
Editor: Keith Wiebe and Noel Gollehon, pp. 183-191

Chapter 20

U.S. ORGANIC AGRICULTURE

Catherine Greene

U.S. farmland under organic management has grown steadily for the last decade as farmers strive to meet consumer demand. By 2003, the United States had over 2 million acres of certified organic crops and pasture. USDA implemented an organic regulatory program in 2002, and Federal research and education activities have also emerged.

INTRODUCTION

Farmers have been developing organic farming systems in the United States since the 1940s, and organic markets have emerged and expanded greatly since then. USDA implemented national standards for organic production and processing in October 2002, following more than a decade of development, and the new uniform standards have facilitated further growth in the organic farm sector. USDA's organic standards incorporate cultural, biological, and mechanical practices that foster cycling of resources, ecological balance, and protection of biodiversity—practices that have evolved over the last half-century.

An increasing number of U.S. farmers are adopting these systems in order to lower input costs, conserve nonrenewable resources, capture high-value markets, and boost farm income. Despite the time, costs, and effort required to meet these stringent requirements, USDA estimates that farmers and ranchers added more than a million acres of certified organic land for major crops and pasture between 1995 and 2003, doubling organic pasture and more than doubling organic cropland for major crops. Total certified organic cropland and pasture now encompasses 2.2 million acres in 49 States. Organic livestock, which require access to organic pasture, have grown more numerous since USDA lifted restrictions on organic meat labeling in late 1999. Food crops and other animal foods (eggs and dairy) are regulated by the Food and Drug Administration and were allowed to carry an organic label throughout the 1990s.

Consumer demand for organically produced goods has risen for over a decade, providing market incentives for U.S. farmers across a broad range of products. Organic products are now available in nearly 20,000 natural food stores and nearly 3 of 4 conventional grocery stores. Organic sales account for approximately 2 percent of total U.S. food sales, according to recent industry statistics. Farmers' markets and other direct-market venues have also grown in number over the last decade, and are especially popular among organic producers. Organic farmers are also finding ways to capture a larger segment of the consumer food dollar through on-farm processing, producer marketing cooperatives, and new forms of direct marketing, including agricultural subscription services.

U.S. Organic Production Standards

Organic farming systems rely on ecologically based practices, such as biological pest management and composting; virtually exclude the use of synthetic chemicals, antibiotics, and hormones in crop production; and prohibit the use of antibiotics and hormones in livestock production. Under organic farming systems, the fundamental components and natural processes of ecosystems—such as soil organism activities, nutrient cycling, and species distribution and competition—are used as farm management tools. For example, crops are rotated, food and shelter are provided for the predators and parasites of crop pests, animal manure and crop residues are cycled, and planting/harvesting dates are carefully timed.

Organic livestock production systems attempt to accommodate an animal's natural nutritional and behavioral requirements, ensuring that dairy cows and other ruminants, for example, have access to pasture. The 2002 USDA livestock standards incorporate requirements for living conditions, pasture and access to the outdoors, feed ration, and health care practices suitable to the needs of the particular species.

The national organic standards address the methods, practices, and substances used in producing and handling crops, livestock, and processed agricultural products. Although specific practices and materials used by organic operations may vary, the standards require every aspect of organic production and handling to comply with the provisions of the Organic Foods Production Act of 1990. Organically produced food cannot be produced using genetic engineering, sewage sludge, or irradiation. These standards include a national list of approved synthetic substances (such as insecticidal soaps and horticultural oils), and prohibited nonsynthetic substances (such as arsenic, strychnine, and tobacco dust) for use in organic production and handling. (See "National Organic Program" on USDA's Agricultural Marketing Service website.)

Adoption of Organic Farming Systems

U.S. farmland under organic management has grown steadily for the last decade as farmers strive to meet consumer demand in both local and national markets. U.S. certified organic crop acreage more than doubled between 1992 and 1997, and doubled again between 1997 and 2003 for many crops. Organic fruit and vegetable crop acreage, along with acreage used for hay and silage crops, expanded steadily between 1997 and 2003. However, most of the acreage increase for organic grain and oilseed crops took place early in this period (Figure 67), and organic soybean acreage has declined substantially since 2001.

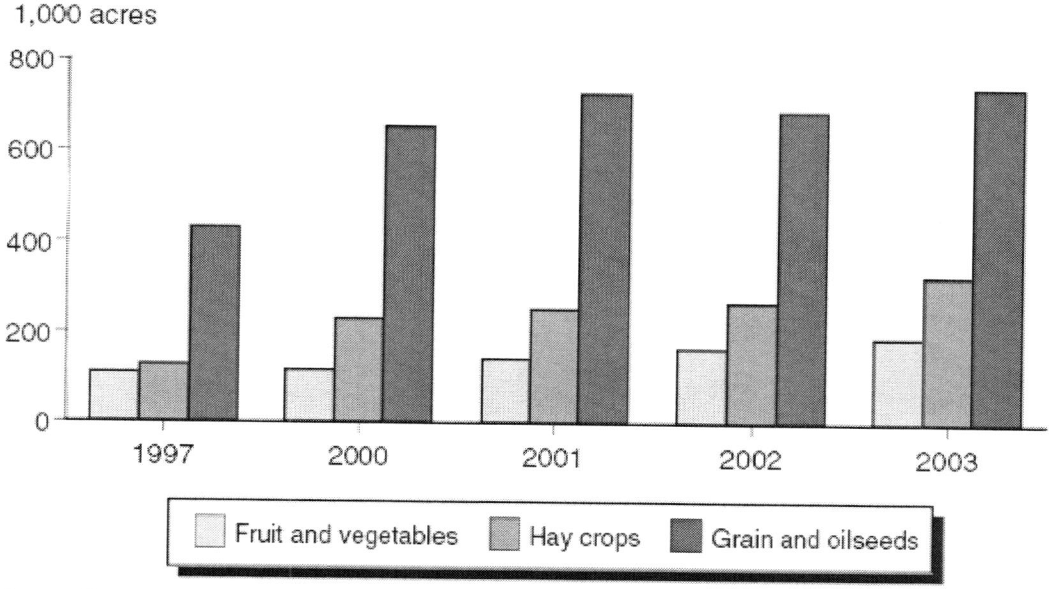

Source: Economic Research Service, USDA.

Figure 67. U.S. certified organic acreage, selected crops, 1997-2003.

Certified organic pasture (including ranchland) declined between 1992 and 1997, but increased 50 percent between 1997 and 2003 after USDA lifted restrictions on organic labeling for meat and poultry. Overall, U.S. farmers and ranchers in 49 States dedicated 2.2 million acres of cropland and pasture to organic production systems in 2003 (Table 31). Many crop/livestock sectors and most States showed strong growth between 2002 and 2003. While certified organic cropland accounted for only 0.4 percent of U.S. cropland in 2001, the share is much higher in some crops, such as fruits (over 2 percent), and vegetables (over 4 percent).

California was the leading State in certified organic cropland in 2003, with nearly 180,000 acres, mostly used for fruit and vegetable production. North Dakota followed with nearly 130,000 acres of cropland, mostly for wheat, soybeans, and other field crops. Minnesota, Montana, Colorado, and Iowa were other top States.

Nearly 40 States had certified pasture and rangeland in 2003, most with under 20,000 acres, although several States had over 100,000 acres and Texas had over 250,000 acres. The number of certified organic beef cows, milk cows, hogs, pigs, sheep, and lambs was up more than five-fold since 1997, and up 15 percent between 2002 and 2003. Dairy has been one of the fastest growing segments of the organic foods industry during this period, and milk cows accounted for over half of the certified livestock animals. Poultry animals raised under certified organic management—including layer hens, broilers, and turkeys—showed even higher levels of growth during 1997-2003.

California had more certified operations than any other State, with just over 1,900 operations in 2003, up 28 percent from the previous year (Figure 68). Wisconsin, Washington, Iowa, Minnesota, New York, Ohio, Vermont, Oregon, and Maine rounded out the top 10. Many of these States have a high proportion of farms with fruits and vegetables and other specialty crops. Also, some of these States, particularly in the Northeast, have relatively little cropland but a large concentration of market gardeners.

Table 31. U.S. certified organic farmland acreage, livestock numbers, and farm operations, 1997 and 2003

Item	1992	1997	2003	Change 1992-1997	Change 1997-2003
				Percent	
U.S. certified farmland:					
Total	935,450	1,346,558	2,196,874	44	63
Pasture/rangeland	532,050	496,385	745,273	-7	50
Cropland	403,400	850,173	1,451,601	111	71
		Number			
U.S. certified animals:					
Livestock--					
Beef cows	6,796	4,429	27,285	-35	516
Milk cows	2,265	12,897	74,435	469	477
Hogs and pigs	1,365	482	6,564	-65	1,262
Sheep and lambs	1,221	705	4,561	-42	547
Total livestock[1]	11,647	18,513	124,346	59	572
Poultry--					
Layer hens	43,981	537,826	1,591,181	1,123	196
Broilers	17,382	38,285	6,301,014	120	16,358
Turkeys	--	750	217,353	--	28,880
Total poultry[2]	61,363	798,250	8,780,152	1,201	1,000
Total certified operations*	3,587	5,021	8,035	40	60

[1]Total livestock includes other and unclassified animals.

[2]Total poultry includes other and unclassified animals.

*Number does not include subcontracted organic operations. Numbers may not add due to rounding.

Source: Economic Research Service, USDA.

Although consumer demand for organic foods is expected to continue growing rapidly in the United States and other major markets, the competition for these markets is likely to increase considerably. Since 2002, USDA has accredited over 40 organizations in foreign countries, as well as approximately 50 groups in the United States, to certify producers and handlers. USDA's Foreign Agricultural Service estimates that the value of U.S. organic exports in 2002 was between $125 million and $250 million, while the value of U.S. organic imports was between $1.0 and $1.5 billion (U.S. Department of Agriculture, 2005). Cotton and soybeans are among the U.S. organic crops that have declined since 2001, despite the growth in retail sales of cotton and soy-based products. Import competition has likely played a role in the decline.

ECONOMIC CHARACTERISTICS OF ORGANIC SYSTEMS

The rapid increase in organic crop and livestock production reflects the increase in consumer demand for organically produced food—20 percent or more per year throughout the 1990s. According to industry data, retail sales of organic food reached $10.4 billion in 2003, up 20.4 percent from the previous year, and accounting for nearly 2 percent of U.S. food sales.

Farmgate and retail price data, collected by private groups, have indicated substantial organic premiums for fruits, vegetables, and milk over the last decade, and recent government data show similar premiums at the wholesale level. Organic grain and soybean crops also enjoyed substantial price premiums during the 1990s, exceeding 50 percent for corn, soybeans, wheat, and oats during 1993-99, and continue to carry a substantial premium (Streff and Dobbs, 2004).

A number of studies have been conducted on the motivations of consumers who purchase organic foods, such as perceived health attributes and concern about pesticide residues and the environment (Dimitri and Greene, 2002). Potential benefits from organic farming systems include improved soil tilth and productivity, lower energy use, and reduced use of pesticides (USDA, 1980; Smolik et al., 1993), and researchers are beginning to compare differences in the nutritive value of the foods produced from these systems as well (Gold, 2000).

A growing number of studies in the United States have examined the yields, input costs, profitability, managerial requirements, and other economic characteristics of organic farming. A 1990 review of the U.S. literature at Cornell concluded that the "variation within organic and conventional farming systems is likely as large as the differences between the two systems," and found mixed results in the comparisons for most characteristics (see Part 4 for more information on conventional systems).

Several USDA and university studies during the 1990s in California, Ohio, and Texas indicated that organic price premiums are necessary to give organic farming systems comparable or higher whole-farm profits than conventional systems, particularly for crops like processed tomatoes and cotton. A Henry A. Wallace Institute of Alternative Agriculture review of university-based comparative studies in the 1980s and early 1990s on Midwestern organic grain and soybean production found organic systems needed price premiums to be more profitable than conventional systems (Welsh, 1999). Several of these studies, however, found that organic grain and soybean production could be as profitable even without price premiums due to higher yields in drier areas or periods, lower input costs, or higher revenue from the mix of crops used in the system. Other recent studies have also found that some organic systems may be more profitable than conventional systems, even without price premiums (Swezey et al., 1994, Reganold et al., 2001).

Net returns to both conventional and organic production systems vary with factors such as soil type, climate, and proximity to markets, and help explain the wide variation in economic performance within each system. Factors not captured in standard profit calculations—such as convenience, longer-term planning horizons, and environmental ethics—can motivate rational adoption of a particular practice or farming system. Our understanding of the factors influencing net returns to organic farming systems remains imperfect.

Economic research on organic farming has tended to focus narrowly on profitability (Fox et al., 1991), but land-grant universities and others are increasingly examining the long-term economics of organic systems through replicated field trial research and a multidisciplinary systems approach (Table 32).

According to the Organic Farming Research Foundation, 18 States had land-grant institutions with research acres under certified organic management in 2003, up from 6 States in 2001. Organic farming systems trials—in experiment stations and on-farm settings—seek to answer basic research questions about yields, profitability, and environmental impacts, as well as to address farmer-defined management and production obstacles to adoption of organic production systems.

USDA's national standards do not restrict additional eco-labeling of organic products, and some organic certifiers are developing standards on social aspects of agricultural production and food distribution—such as fair trade and local sourcing. The Florida Certified Organic Growers and Consumers organization, for example, recently developed a partnership program for food retailers and restaurants in North Florida to certify their level of commitment to local food sourcing (FOG/QCS, 2003). Most coffee sold in the United States that is certified as fair trade (indicating that farmers receive a fair price) also has a separate organic certification. Some certification groups are trying to improve the efficiency of their efforts by integrating these programs.

FEDERAL POLICY INITIATIVES

Government research and policy initiatives often play a key role in the adoption of new farming technologies and systems. Worldwide, adoption levels for organic farming systems are currently the highest in European Union countries. Governments there have been developing consumer education initiatives and providing direct financial support to producers for conversion since the late 1980s to capture environmental benefits and support rural development.

Table 32.—Examples of U.S. multidisciplinary, long-term research projects with organic trials

Project	Date established	Farming system/Commodity focus
University of Nebraska-Lincoln *Long-Term Experiment Trials*	1975	Compare conventional and organic systems (Rotations include corn, wheat, and soybean)
Rodale Institute—Kutztown, PA *Farming Systems Trial TM*	1981	Examine the transition process from conventional to organic farming (corn and soybeans)
University of California-Davis *Sustainable Agriculture Farming Systems Project*	1988	Compare conventional, low-input and organic systems; evaluate conservation tillage in these systems (tomato, safflower, bean, corn)
Iowa State Univ.—Leopold Center *Neely-Kinyon Long-Term Agroecological Research*	1988	Compare conventional and organic systems (corn, soybeans and alfalfa)
University of Minnesota-Lamberton Experiment Station *Elwell Agroecology Farm*	1989	Compare conventional and organic systems (corn, soybeans, alfalfa and oats)
Michigan Agricultural Experiment Station *Living Field Laboratory*	1993	Compare conventional, organic and other systems (corn, soybeans, and wheat)
USDA Agricultural Research Center-Beltsville, MD *Farming Systems Project*	1993	Compare organic systems typical in the mid- Atlantic region (corn and soybeans)
West Virginia University (WVU) *Horticulture Farm Project*	1999	Evaluate organic systems on the entire Horticulture Farm (market garden and field crop/livestock systems)
North Carolina State University *Farming Systems Trial*	2001	Compare conventional and transitional organic systems (grains, livestock and woodlots)
Ohio State University *John Hirzel Sustainable Agriculture*	2001	Compare conventional, no-till, and high-and lowinput organic systems (soybeans, corn, wheat, and

Research and Education Site	vegetables)

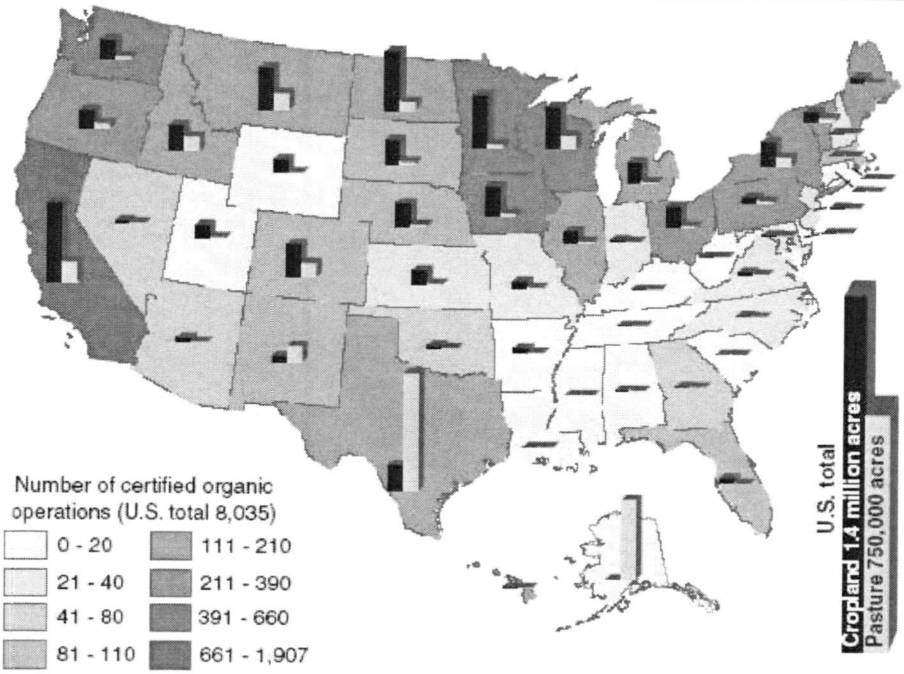

Source: Economic Research Service, USDA.

Figure 68. Certified organic acreage and operations, 2003.

State and Federal support for organic farmers and handlers is also beginning to emerge in the United States. Minnesota and Iowa, for example, began subsidizing conversion to organic farming systems in the late 1990s as a way to capture the environmental benefits of these systems. In 2003, the National Association of State Departments of Agriculture released a policy statement on organic agriculture expressing support for a wide range of activities that would expand public-sector organic research and education and provide technical assistance to organic and transitional farmers.

USDA agencies have started or expanded programs on organic agriculture during the last several years, and the Farm Security and Rural Investment Act of 2002 (the 2002 Farm Act) contains several first-time research and technical assistance provisions to directly assist organic producers. Recent programs and initiatives include:

- *Certification Cost-Share Support.* In 2001, USDA established a certification cost-share program to help farmers defray certification costs in 15 States. The 2002 Farm Act allocated $5 million in cost-share assistance funds for this program and expanded eligibility to growers and handlers in all States.
- *Research and Technical Assistance.* The 2002 Farm Act contains an Organic Agriculture Research and Extension Initiative that authorizes $3 million per year in new mandatory appropriations in fiscal years 2003-07. These funds are being used to administer competitive research grants focused on organic agriculture production, breeding, and processing methods, as well as the marketing and policy constraints.

- *Conservation Initiatives.* The 2002 Farm Act provided funding for the Conservation Security Program, which provides payments to producers for adopting or maintaining land management and conservation practices to address resource concerns. This new program may interest organic farmers who commonly adopt these types of practices as part of their organic farming systems. (See Part 5 for a general discussion of Federal conservation initiatives and for a discussion of working-land programs.)
- *Exemptions From Marketing Assessments.* Another provision in the 2002 Farm Act specifies that certified organic producers who produce and market only organic products and do not produce any conventional or nonorganic products will now be exempt from paying an assessment under any commodity promotion law. In December 2003, USDA published a proposed rule to exempt producers from paying assessments associated with its marketing order programs, and in January 2005 a similar proposed rule was published to cover exemptions for its research and promotion programs.
- *Export Promotion, Crop Insurance, and Other Initiatives.* USDA's Risk Management Agency has provided insurance coverage for organic farming practices as good farming practices by written agreement since 2001, and made organic farmers eligible for a wider range of coverage options in 2004. USDA's Foreign Agricultural Service (FAS) is helping design protocols for working with foreign nations to keep organic trade moving as more countries develop organic standards.

REFERENCES

[1] Dimitri, Carolyn, and Catherine Greene (2002). *Recent Growth Patterns in the U.S. Organic Foods Market*, AIB-777, U.S. Dept. Agr., Econ. Res. Serv., Sept.

[2] FOG/QCS (2003). *Certification Manual of North Florida Local Food Partnership, 2003-2004*, Florida Organic Growers & Consumers/Quality Certification Services. Gainesville, FL.

[3] Fox, Glenn, Alfons Weersink, Ghulam Sarwar, Scott Duff, and Bill Deen (1991). "Comparative Economics of Alternative Agricultural Production Systems: A Review," *Northeastern Journal of Agricultural and Resource Economics,* Vol. 20.

[4] Gold, Mary (2000). *"Organically Produced Foods: Nutritive Content,"* Special Reference Briefs Series SRB 2000-03, Alternative Farming Systems Information Center, National Agricultural Library, U.S. Dept. Agr., Agricultural Research Service.

[5] Greene, Catherine, and Amy Kremen (2003). *U.S. Organic Farming in 2000-2001: Adoption of Certified Systems*, AIB-780, U.S. Dept., Agr., Econ. Res. Serv., June.

[6] National Association of State Departments of Agriculture. 2003. *Organic Agriculture*, NASDA Policy Statement. Sept. 11.

[7] Reganold, J.P., J.D. Glover, P.K. Andrews, and H.R. Hinman (2001). "Sustainability of three apple production systems," *Nature*, 410: 926-930.

[8] Smolik, James D., Thomas L. Dobbs, Diane H. Rickerl, Leon J. Wrage, George W. Buchenau, and Thomas A. Machacek (1993). *Agronomic, Economic, and Ecological Relationships in Alternative (Organic), Conventional, and Reduced-Till Farming Systems*, B718, Agricultural Experiment Station, South Dakota State University, Sept.

[9] Streff, Nicholas, and Thomas L. Dobbs (2004). *'Organic' and 'Conventional' Grain and Soybean Prices in the Northern Great Plains and Upper Midwest: 1995 through 2003,* South Dakota State University Economics Pamphlet 2004-1, Brookings, SD, June.

[10] Swezey, Sean, Jim Rider, Matthew Werner, Marc Buchanan, Jan Allison, and Stephen Gliessman (1994). "Granny Smith conversions to organic show early success," *California Agriculture*, Vol. 48, No. 6.

[11] U.S. Department of Agriculture (2000). "National Organic Program, Final Rule," *Federal Register* 7 CFR Part 205, Dec. 21.

[12] U.S. Department of Agriculture (2005). "U.S. Market Profile for Organic Food Products," Commodity and Marketing Programs—Processed Products Division, International Strategic Marketing Group, Feb. 22.

[13] U.S. Department of Agriculture, USDA Study Team on Organic Farming (1980). *Report and Recommendations on Organic Farming*, U.S. GPO No. 620-220/3641, July.

[14] Welsh, Rick (1999). *The Economics of Organic Grain and Soybean Production in the Midwestern United States*, Policy Studies Report No. 13, Henry A. Wallace Institute for Alternative Agriculture, May.

PART 5: CONSERVATION AND ENVIRONMENTAL POLICIES

In: Agricultural Resources and Environmental Indicators ISBN 978-1-60021-467-7
Editor: Keith Wiebe and Noel Gollehon, pp. 195-202

Chapter 21

CONSERVATION POLICY OVERVIEW

Roger Claassen and Marc Ribaudo

USDA implements a broad range of conservation programs intended to protect natural resources and the environment. The Farm Security and Rural Investment Act of 2002 sharply expanded funding for conservation programs, focusing much of the increase on programs for working agricultural lands, e.g., cropland and grazing land.

INTRODUCTION

Some farming practices (excess fertilization and manure, for example) can degrade our Nation's natural resources while others (such as land reservation for wildlife) can enhance our natural heritage. Policymakers have been devoting more attention and funding to conservation programs that support environmental enhancement and reduce the potential for environmental harm. Until 2002, the bulk of conservation funds went toward land retirement: paying farmers to remove environmentally sensitive land from crop production for a specified time. With the 2002 Farm Security and Rural Investment Act (2002 Farm Act), policymakers substantially increased conservation funding, especially on lands used for crop production and grazing (Figure 69).

By 2007—if authorized levels are realized—conservation funding will be double the level under the previous farm bill (1996-2001), with about two-thirds of the new funds going to programs emphasizing conservation on working lands.

USDA CONSERVATION PROGRAMS IN RELATION TO ALL ENVIRONMENTAL EXPENDITURES

Agricultural conservation programs are part of a larger Federal effort to protect and preserve natural resources (Table 33). Conservation and land management efforts include agriculture, but also encompass programs of the Forest Service, Bureau of Land Management, Fish and Wildlife Service, and other Federal agencies (see Chapter 27, "Federal Laws

Protecting Environmental Quality"). Funding for water resource programs, recreational services, and pollution control/abatement activities also come under the general rubric of natural resources. Agricultural conservation spending was about 17 percent of the $32.7 billion in Federal spending for natural resources in fiscal year 2004.

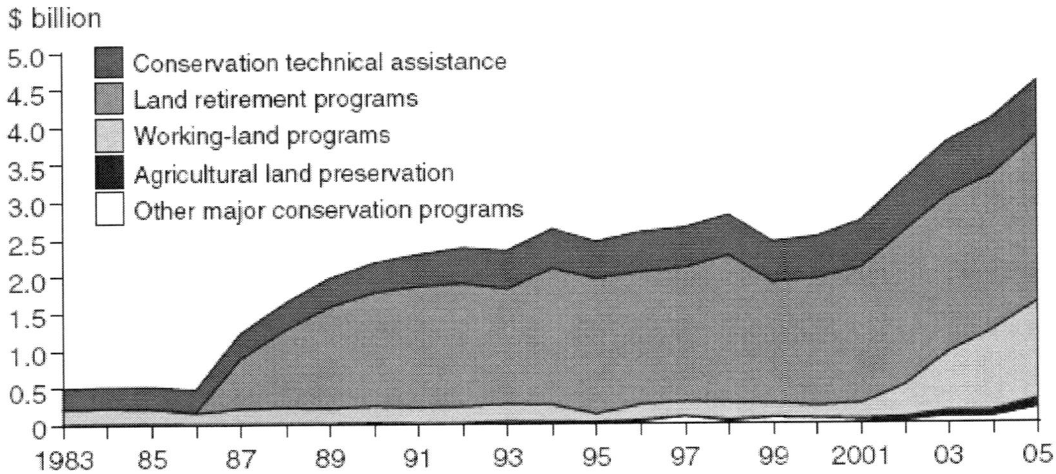

Source: ERS analysis of Office of Budget and Program Analysis data.

Figure 69. Trends in USDA conservation expenditures, 1983-2005.

USDA CONSERVATION PROGRAMS: A PORTFOLIO APPROACH

USDA conservation programs have traditionally used voluntary approaches to natural resource issues. These approaches can avoid the inherent difficulties in regulating nonpoint sources of pollution and can minimize economic harm to farmers by educating them and providing them with incentives to willingly improve production practices. In passing the 2002 Farm Act, Congress reaffirmed a preference for addressing natural resource problems on private land through a consolidated set of financial assistance programs supported by research and education. In a notable exception, Conservation Compliance (see Chapter 23, "Compliance Provisions for Soil and Water Conservation"), which requires wetland conservation and soil conservation on highly erodible cropland for producers receiving Federal farm program payments, was continued. USDA programs now, more than ever before, offer producers a range of options for assistance with conservation efforts, among them (see Table 34):

- Land retirement programs generally remove land from agricultural production for a long period (at least 10 years) or, in some cases, permanently.
- Working-land programs provide technical and financial assistance to farmers who install or maintain conservation practices on land in production.
- Agricultural land preservation programs purchase rights to certain land uses, such as development, in order to maintain land in agricultural use.

- USDA provides, through Conservation Technical Assistance, ongoing technical assistance to agricultural producers who seek to improve the environmental performance of their farms.

Table 33. Federal natural resources expenditures (budget authority), FY 2004

Subfunction and agency/activity	Discretionary programs	Mandatory programs	Total
		$ million	
Water resources			
Corps of Engineers	4,424		4,424
Bureau of Reclamation	906		906
Watershed, flood prevention, and other	357		357
Conservation and land management			
Forest Service	5,116		5,116
Bureau of Land Management	1,776		1,776
Conservation of agricultural lands	900	4,598	5,498
Fish and Wildlife Service	1,222	1,222	
Other conservation and land management programs	754		754
Recreational services			
Operation of recreational resources	2,340	956	3,296
Other recreational resource activities	28		28
Pollution control and abatement			
Regulatory, enforcement, and research programs	3,188		3,188
State and tribal assistance programs	3,877		3,877
Hazardous substances superfund	1,258	85	1,343
Other pollution control and abatement activities	164		164
Other natural resources			
National Oceanic & Atomospheric Administration	3,738		3,738
Other natural resource program activities	1,101		1,101
Fee and mandatory programs	14		14
Total gross budget authority	31,149	5,653	36,802
Offsetting receipts	-15	-4,065	-4,080
Net budget authority	31,134	1,588	32,722

Source: Office of Management and Budget (OMB).

This "portfolio" approach to conservation policy provides the flexibility needed to address agri-environmental issues. Most producers—regardless of their agri-environmental problems, resource settings, and the size and management structure of their operation—have options for receiving Federal assistance for conservation.

Smaller operations—those with sales of less than $250,000 per year—produce roughly one-third of U.S. agricultural output but include nearly three-quarters of all producer-owned land. Operators of these farms often receive a larger share of their household income from land retirement payments and non-farm sources than from the sale of agricultural products.

Table 34. Funding for major USDA conservation programs, 2002-2005

Program type and program	2002	2003	2004	2005[1]
		$ million		
Land Retirement				
Conservation Reserve Program	1,785	1,789	1,799	1,937
Wetlands Reserve Program	284	309	285	268
Working Land				
Environmental Quality Incentives Program	390	331	904	995
Ground and surface water	25	54	66	54
Klamath Basin	2	12	19	9
Wildlife Habitat Incentives Program	15	24	38	47
Conservation Security Program			41	202
Agricultural Land Preservation				
Farm and Ranch Land Protection Program	51	78	91	112
Other				
Grassland Reserve Program		39	55	128
Emergency Conservation Program	32	47	23	80
Conservation Technical Assistance	679	716	742	720
Total, major conservation programs	3,263	3,398	4,062	4,552

[1]Estimated

Source: ERS analysis of Office of Budget and Program Analysis data.

Larger farms, on the other hand, produce two-thirds of U.S. agricultural output while accounting for only one-fourth of the land. These farms are generally more commercially oriented and their operators receive most of their household income from farm sources. The 2002 Act's increased funding for conservation on working lands, along with a greater focus on livestock operations and relaxation of conservation payment limitations, is expected to raise conservation participation by larger farms.

Expanding Conservation on Working Lands

Authorized funding for the **Environmental Quality Incentives Program (EQIP)**, the major working lands program (see Chapter 24, "Working-Land Conservation Programs"), jumped five-fold with the 2002 Farm Act, approaching $5.8 billion for 2002-07. Of the $3.3 billion authorized for FY2002-2005, $3 billion (91 percent) has been made available.

Through EQIP, crop and livestock producers can get information and technical/financial assistance in designing and implementing conservation practices (structural or land management) on their land. In response to new (2003) Clean Water Act regulations on animal feeding operations, EQIP now provides more incentives for livestock producers to participate. At least 60 percent of the program's funding is targeted for livestock producers, up from 50 percent in the 1996 Farm Act. Limits on the size of participating livestock operations have been removed, and maximum payment levels per year have been increased. EQIP will also put greater emphasis on water conservation. A new, separate fund for ground- and surface-water conservation activities was established within EQIP, as well as a special fund for water conservation in the Klamath Basin in California and Oregon.

The 2002 Farm Act also authorized a new working-land program: the **Conservation Security Program (CSP)**. Like EQIP, CSP encourages producers to address resource concerns such as soil quality, water quality, or wildlife habitat on working land. The differences in these two programs, however, are greater than the similarities. Unlike other conservation programs, CSP was approved as an entitlement program, meaning that eligible producers who meet program requirements can be enrolled at the producers'option, as in ongoing commodity programs. Before CSP was implemented, however, Congress limited CSP funding, making limitations on enrollment necessary. For FY 2004, $41 million was available for CSP, and $202 million more was available in FY 2005.

Unlike EQIP, CSP requires a substantial level of environmental stewardship before producers become eligible for enrollment. Soil quality and water quality must be addressed (to standards set by USDA's Natural Resources Conservation Service (NRCS)) before land can be enrolled in CSP. CSP rewards these and other past conservation efforts through stewardship payments. CSP also encourages whole-farm conservation by offering higher stewardship payments to producers who undertake farmwide conservation.

CSP also funds "enhancements,"which are directed, in part, toward encouraging producers to go beyond basic conservation effort encouraged by more traditional programs like EQIP (e.g., to reduce erosion below the soil loss tolerance—the traditional standard for soil erosion control). While many livestock-related practices can be eligible for CSP, the focus is on land-based practices; livestock waste management structures and handling equipment are specifically excluded. Finally, CSP is a national program, but is available only in selected watersheds for any given signup. Part of the NRCS strategy is to make all 2,119 U.S. watersheds eligible for enrollment at least once over an 8-year period.

The **Wildlife Habitat Incentives Program (WHIP)** provides cost-sharing to landowners and producers to develop and improve wildlife habitat. WHIP funding rose from just over $62 million during 1996-2001 to $360 million over FY 2002-07. For FY 2002-2005, WHIP has received funding of $129 million, 68 percent of the $190 million authorized for that period.

Accompanying the large increase in working-land program funding was a more subtle change in the way funds are awarded through these programs—changes that may reduce environmental cost effectiveness. In EQIP, for example, the 2002 Farm Act eliminated the use of conservation priority areas, which focused the program's effort in areas of highest environmental need. The Act also eliminated "bidding down," which allowed producers to increase their chance of enrollment by offering to take a smaller payment, reducing the cost of contracts and thereby stretching the program budget. For contract offers with comparable environmental values, the Secretary of Agriculture cannot assign higher priority to an application based only on a lower bid from the operator. NRCS may consider costs in ranking potential participants, even though bidding down is no longer allowed.

While the expansion of conservation on working lands offers significant benefits, implementing it may pose additional challenges. Payments for a broad range of conservation practices on working land are now available to a wider range of producers than ever before, expanding the importance of both conservation planning and monitoring of practice implementation and maintenance. This is particularly true for some conservation management practices, such as crop nutrient management, which are less visible and thus more difficult to monitor than changes in tillage or contour cropping. Multiple conservation programs for working lands could also make it more difficult for programs to work together seamlessly and avoid duplication.

Producers participating in new and newly expanded conservation programs will need conservation planning services and technical assistance. To help handle the increased workload, the 2002 Farm Security and Rural Investment (FSRI) Act included funding for certification of third-party technical service providers to supplement NRCS field staff.

Land Retirement

Land retirement programs (see Chapter 22, "Land Retirement Programs") pay producers to remove land from crop production. In exchange for retiring land, producers receive rental or easement payments, plus cost sharing and technical assistance to help establish and maintain permanent cover. Economic use of the land is limited.

Land retirement dominated Federal agricultural conservation spending between 1985 and 2002 and continues to be the largest single component of agricultural conservation spending (Figure 69, Table 34). In FY 2000, 90 percent of cash conservation payments made directly to producers were associated with land retirement. Between 1985 and 2000, roughly 50 percent of all USDA conservation spending was for land retirement. (USDA conservation spending also includes cost sharing and technical assistance for non-land retirement activities, public works, and a range of other administrative, data collection, and research activities.)

While the expansion of working-land programs was the big story in the 2002 Farm Act, land retirement programs also grew, particularly for wetland restoration. While wetland restoration accounts for about 3 percent of current land retirement, 40 percent or more of the authorized increase in retired acreage may be devoted to wetlands restoration. The shift toward wetlands restoration is significant because of the high environmental benefits per acre provided by wetlands relative to other types of land cover (e.g., grass).

The 2002 Farm Act increased the **Conservation Reserve Program (CRP)** acreage cap from 36.4 million acres to 39.2 million acres (Table 35).

The **Wetlands Reserve Program (WRP)** acreage cap was more than doubled from 1.075 million acres to 2.275 million. The Farm Act also required enrollment of 250,000 new acres per year. In addition to the 1.2 million acres added to WRP, the CRP routinely enrolls farmed wetlands that are restored to wetland condition. Up to 500,000 acres of the 2.8-million-acre rise in the CRP could be earmarked for restoration of currently farmed wetlands. At the end of FY 2004—about halfway through the period covered by the 2002 Farm Act—total CRP enrollment stood at 34.9 million acres, while WRP enrollment totaled nearly 1.7 million acres (Table 35).

Agricultural Land Preservation

The **Farm and Ranchland Protection Program (FRPP)** (see Chapter 26, "Farmland Protection Programs") provides funds to State, tribal, or local governments and private organizations to help purchase development rights and keep productive farmland in agricultural use. FRPP received just over $50 million total during 1996-2001. The 2002 Farm

Act authorized funding of $597 million over FY 2002-07. For FY 2002-2005, $352 million was made available, 88 percent of the $400 million authorized.

Table 35. Acreage-limited programs, pre-and post-2002

Program	Pre-2002 acreage cap	New authority, 2002			
		Wetlands	Other acreage	New acreage cap	Enrolled (through FY 2004)
			Million acres		
Conservation Reserve Program	36.4	≥ 0.5	≤ 2.3	39.2	34.9
Wetlands Reserve Program	1.1	1.2	0.0	2.3	1.7

Source: ERS analysis of NRCS and Farm Service Agency data.

Other Conservation Programs

The **Grassland Reserve Program (GRP)** (see Chapter 25, "Conservation on Private Grazing Lands") is designed to improve and conserve native-grass grazing lands through long-term rental agreements (10, 15, 20, or 30 years) and 30-year or permanent easements. While normal haying and grazing activities are allowed, producers and landowners are required to (1) restore and maintain appropriate grasses, forbs, and shrubs; (2) address all relevant resource concerns (e.g., soil erosion); and (3) refrain from converting the land for crop production, development, or other uses. For rental agreements, annual rental payments equal (up to) 75 percent of grazing value. Permanent easements are to be purchased at fair market value, less grazing value, while 30-year easements are to be purchased at 30 percent of the value of a permanent easement. Cost-sharing is provided for up to 75-90 percent of the restoration and maintenance costs, depending on the type of grassland. GRP enrollment is limited to 2 million acres of grassland. Funding of up to $254 million is authorized over the 6-year life of the Farm Act. During FY 2003-2005, $177 million was made available to producers through GRP.

The **Emergency Conservation Program (ECP)** helps farmers to rehabilitate farmland damaged by natural disasters. In particular, it addresses problems that, if left untreated, would (1) impair or endanger the land, (2) reduce the productive capacity of the land, (3) be so costly to rehabilitate that Federal assistance would be required to return the land to productive agricultural use, or (4) represent damage that is unlikely to recur in the same area.

Watershed Programs and RC&D

A final group of USDA programs provides conservation protection for watersheds and includes Resource Conservation & Development (RC&D) (Table 36). Watershed protection programs generally assist local communities with flood protection, water supply, and water quality.

Table 36. Watershed programs and RC&D, FY 2002-2005

Program	2002	2003	2004	2005
			$ million	
Watershed and flood prevention operations	210	121	86	75
Watershed surveys and planning	11	11	10	7
Watershed Rehabilitation Program	10	29	30	27
Resource Conservation & Development	48	50	52	51
Total	279	211	178	160

Source: USDA Office of Budget and Program Analysis.

In: Agricultural Resources and Environmental Indicators ISBN 978-1-60021-467-7
Editor: Keith Wiebe and Noel Gollehon, pp. 203-212

Chapter 22

USDA LAND RETIREMENT PROGRAMS

Daniel Hellerstein

USDA's primary land retirement programs are the Conservation Reserve Program (CRP) and the Wetlands Reserve Program (WRP). Started in 1986, the CRP has retired over 34 million acres of environmentally sensitive crop-land under 10-to 15-year contracts. The WRP, started in 1992, protects over 1.6 million acres of wetlands, primarily using permanent easements.

INTRODUCTION

In 2004, USDA's land retirement programs accounted for over half of all the Department's conservation expenditures (See Part 5 for an overview of conservation programs.). Under these programs, the Government offers rental payments and other incentives to farm owners and operators, who convert land from agricultural production to land covers deemed more environmentally beneficial. In 2004, USDA spent over $1.6 billion on the Conservation Reserve Program (CRP) to retire over 34 million acres of cropland. In addition, the $280 million spent on the Wetlands Reserve Program (WRP) increased protected wetland acreage to over 1.6 million acres. Although the Farm Security and Rural Investment Act of 2002 signaled a shift toward working lands programs, land retirement will continue to be important. In this chapter, we review the trends, status, and challenges facing both the CRP and the WRP.

THE CONSERVATION RESERVE PROGRAM

The Conservation Reserve Program (CRP) was established by the Food Security Act of 1985 and began enrolling farmland in 1986. The program uses contracts with agricultural producers and landowners to retire highly erodible and environmentally sensitive cropland and pasture from production for 10-15 years. Enrolled land is planted to grasses, trees, and other cover, thereby reducing erosion and water pollution and providing other environmental benefits (as well as reducing the supply of agricultural commodities).

Enrollment in CRP increased rapidly once the program got underway (Figure 70), with nearly all eligible applicants accepted. Approximately 34 million acres were enrolled during the first 9 signups (between 1986 and 1989).[1] In these early years, CRP eligibility was limited to about 100 million acres of land with highly erodible soils, with per-acre payments based on a regional average of cropland rental rates (along with half the cost of establishing permanent cover).[2]

The CRP was not the first farmland retirement program operated by the Federal Government, nor was it the only land diversion program operating at the time of its enactment. The Soil Bank Program, established in 1956, expired in the early 1970s. Furthermore, annual paid land diversion and Acreage Reduction Program (ARP) requirements continued through 1995. In fact, diverted acres outnumbered CRP enrollment until 1990 (Figure 70). However, these earlier land diversion programs focused on supply control and did not require environmental/habitat management. The primary goal of the CRP in the years immediately following its creation was to reduce soil erosion on highly erodible cropland (Osborn and Heimlich, 1994).[3]

The Food, Agriculture, Conservation, and Trade Act of 1990 expanded eligibility for CRP beyond highly erodible land. The 240 million acres of eligible land included several "Conservation Priority Areas" (the Chesapeake Bay, Long Island Sound, and Great Lakes watersheds), State water quality priority areas, and smaller plots of land adopting high-priority conservation practices (Barbarika, 2001).

USDA also made two significant changes to program enrollment criteria:

- To account for multiple environmental concerns, an environmental benefits index (EBI) was used to rank offers. The EBI weights a number of different concerns, including water quality, air quality, and soil erodibility (Table 37).
- Maximum allowable rental rates were based on a soil-specific estimate of the rent earned on comparable local cropland. Use of soil-specific maximum rental rates enabled USDA to enroll environmentally sensitive, but highly productive, land into the program.

Following passage of the Federal Agriculture Improvement and Reform Act of 1996, wildlife habitat was added to the EBI. A continuous signup was initiated for acreage devoted to specific conservation practices, such as filter strips, riparian buffers, grassed waterways, field windbreaks, shelterbelts, living snow fences, salt-tolerant vegetation, shallow water

1 Although the original (1985) legislation envisioned the program retiring 40-45 million acres, enrollment authority was capped at 38 million acres in 1992 and reduced to 36.4 million acres in 1996. In 2002, CRP's enrollment authority was increased to 39.2 million acres.

2 The amount of land enrolled in CRP does not necessarily reflect the amount of land removed from production. First, "slippage," the reallocation of lands outside the program (such as pastureland) to cropland uses, may occur. Wu (2000) argues that about 21 acres have been brought into crop production for every 100 retired through CRP. However, more recent studies using the same data have found no evidence of slippage in the CRP (Roberts and Bucholtz, 2004). Second, land enrolled in CRP might have left production even without the program. Lubowski et al. (2003) estimate this to be about 8 percent of CRP acres.

3 It has been argued that, given the financial crisis facing the farm sector in the mid-1980s, curbing farm production and supporting income of CRP participants were equally important program goals (Dicks, 1987).

areas for wildlife, and wellhead protection. In 1997, continuous signups were augmented by the Conservation Reserve Enhancement Program (CREP), a Federal-State partnership designed to encourage farm conservation practices that meet specific State and national conservation and environmental objectives.

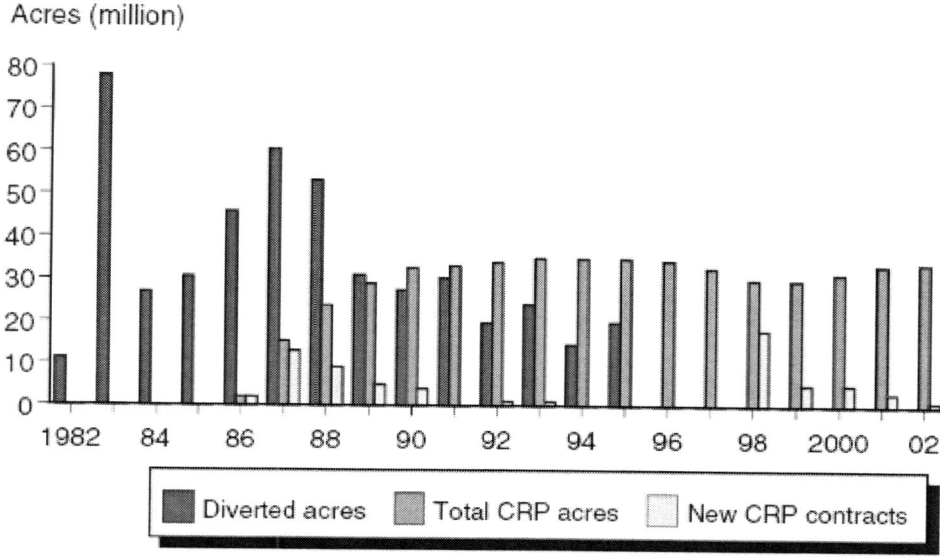

Notes: Diverted acres includes land enrolled in the Soil Bank, and land used to fulfill Acreage Reduction Program requirements.

Source: Farm Service Agency CRP Summary Statistics and U.S. Land Use Summary.

Figure 70. CRP enrollment and other diverted acreage.

Table 37. Assignment of EBI points in the 26th CRP signup

Wildlife 100	Cover (introduced grass, native grass, trees) 50	Priority zones 30	Wildlife enhancement 20
Water quality 100	Within designated State water quality zone 30	Groundwater vulnerability 25	Surface-water vulnerability 45
Erosion 100	Erodibility index 100		
Enduring benefits 50	Enduring benefits (tree plantings, wetland restoration, existing trees, grass seeding) 50		
Air quality 45	Air quality benefits 35	Wind erosion Soils 5	In air quality zones 5
Costs 150	Per acre rent 125 125*(185-bid_amount)/185 (185 is CRP's maximum allowed bid)	No cost-share 10	Bid below maximum Rate 15

With early contracts expiring, signups conducted in 1997 and 1998 enrolled over 22 million acres. Unlike the early signups, competition was keen, with all bids ranked using the EBI. Since the bid process meant that already enrolled lands were not automatically re-enrolled, the distribution of CRP enrollment shifted somewhat during the 1990s. Of the nearly 34 million acres enrolled in 2002, 17 percent represented net additions to county CRP acreage (over the county's 1990 enrollment). And of the nearly 33 million acres enrolled in 1990, 14 percent was dropped from the program by 2002.

Although a roughly equal number of counties gained and lost CRP acreage between 1990 and 2002, there was little redistribution of acreage at the regional level (Sullivan et al., 2004). The Northern Great Plains gained slightly, at the expense of the Heartland (probably due to the lower rental rates requested by Plains bidders) and the Southern Seaboard (where many CRP acres planted in trees were not offered for re-enrollment).

Overall, the CRP started as a program with a soil conservation agenda, in a time when the farm sector was weathering a severe economic downturn. As other stakeholders recognized the potential of this dedicated stream of conservation expenditures, CRP evolved beyond soil conservation, with greater weight given to wildlife habitat, air and water quality, and carbon sequestration.

The Wetlands Reserve Program

The Wetlands Reserve Program (WRP) was established by the Food, Agri-culture, Conservation, and Trade Act of 1990. WRP goals are the restoration of high-risk agricultural land located in, or adjacent to, floodprone areas. The stated emphasis of WRP is to protect, restore, and enhance the functions and values of wetland ecosystems to attain:

- Habitat for migratory birds and wetland-dependent wildlife, including threatened and endangered species,
- Protection and improvement of water quality,
- Attenuation of water flows due to flooding,
- Recharge of ground water,
- Protection and enhancement of open space and aesthetic quality,
- Protection of native flora and fauna contributing to the Nation's natural heritage, and
- Contribution to educational and scientific scholarship.

WRP enrollment began in 1992, with steady increases in subsequent years (Figure 71). When the initial enrollment cap of 1 million acres was met in 2001, the Farm Security and Rural Investment Act of 2002 reauthorized the WRP, increasing the cap to 2.275 million acres. The WRP uses three enrollment schemes: permanent easements, 30-year easements, and 10-year cost-share agreements.

The initial 2 years of enrollment consisted of pilot programs in a limited number of States. WRP has since sought the greatest wetland functions and values, along with optimum wildlife habitat, on every acre enrolled. In pursuing these goals, WRP has undergone some changes. Most importantly, in the earlier years a "walk away" strategy was often used: parcels were allowed to return to their wetland condition with no other intervention. However, this strategy led to poor wetland function. So, a "full restoration" strategy was adopted in the late 1990s. Full restoration implies considerably more site preparation (for example, undoing land leveling). At least 70 percent of each project must be restored to the original natural condition (to the extent practicable). The remaining 30 percent can be restored to "other than natural" conditions.

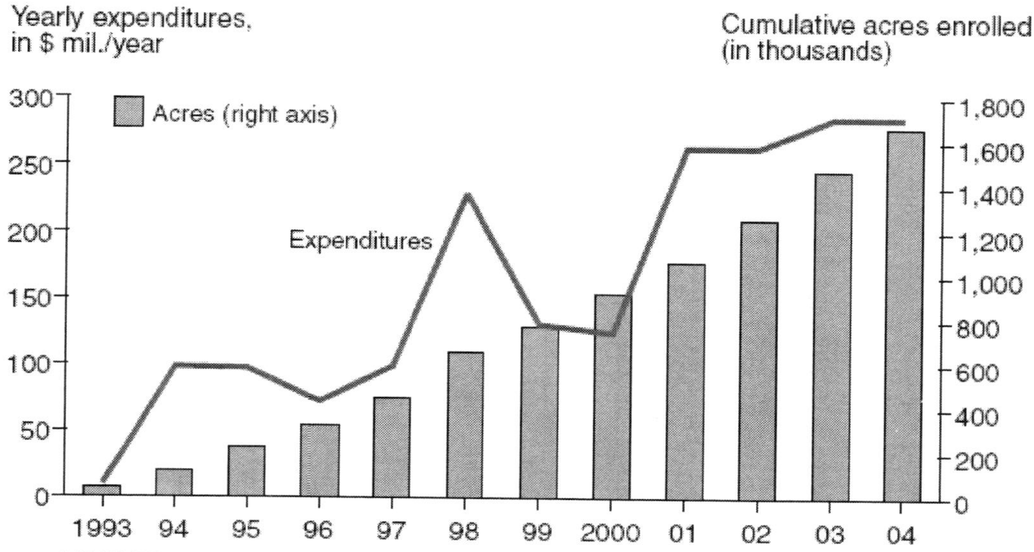

Source: NRCS/USDA.

Figure 71. WRP expenditures and cumulative enrollment.

Current Status of Land Retirement

As of January 2005, the CRP enrolled 34.8 million acres of land at a cost of $1.68 billion per year (average cost of about $45/acre). The bulk of this land was enrolled via "general" signup—about 31.7 million acres (Table 38). The remaining acres are in "continuous" signup, which includes 117,000 acres of farmable wetlands (small non-floodplain wetlands). Most CRP land is in the Northern Great Plains, Prairie Gateway, and Heartland (Figure 72).

As of September 2004, the WRP enrolled 1.6 million acres of land, mostly in permanent easements. Expenditures in 2004 were about $275 million spread over 189,000 acres (an average cost of $1,400 per acre). Average contract size is 194 acres (Table 39). Much of WRP land is in Missouri, Arkansas, Louisiana, Mississippi, Florida, and California (Figure 73).

Challenges

Over their 20-year-plus lifespans, both the CRP and WRP have provided an array of environmental benefits (Table 40). While this suggests that both programs are successful, each faces challenges.

Table 38. CRP status of January 2005

Signup type	Contracts	Farms	Acres	Annual rental payments ($ million)	($/acre)
General[1]	394,767	262,076	31,753,754	1,384	43.59
Continuous					
Non-CREP[2]	234,916	147,616	2,259,265	201	89.11
CREP[3]	40,067	26,775	631,098	76	120.31
Subtotal	274,983	170,448a	2,890,363	277	95.92
Farmable wetland[4]	7,938	6,450	122,803	15	119.12
Total	677,688	397,970a	34,766,920	1,676	48.21

[1]General signup. Held on a more-or-less yearly basis, producers with eligible lands compete nationally for acceptance based on an environmental benefits index (EBI).

[2]Continuous (Non-CREP) signup. Producers with eligible lands may enroll certain high-priority conservation practices, such as filter strips and riparian buffers, at any time during the year without competition. In addition to annual soil rental payments and cost-share assistance, many practices are eligible for additional annual and one-time upfront financial incentives.

[3]Conservation Reserve Enhancement Program (CREP). There are currenty 29 CREP Federal/State partnerships, which implement projects designed to address specific environmental objectives through targeted CRP enrollments. Signup is continuous.

[4]Farmable Wetlands Program (FWP). Producers enroll small non-floodplain wetlands under modified continuous signup provisions.

Source: FSA/USDA.

Source: ERS, based on data from the Farm Service Agency, USDA.

Figure 72. Distribution of CRP lands, 2004.

Table 39. WRP status as of September 2004

Total enrolled acres	1.6 million
Average acres per contract	194
Size of contracts (percent of program acres):	
< 100 acres	61
100 to 500 acres	32
501 to 1,000 acres	5
> 1,000 acres[1]	2
Type of easement (percent of program acreage):	
Permanent	80
30 year	14
10 year[2]	6

[1]Many of the larger projects are the result of multiple landowners enrolling in the program, creating a single large area of land.

[2]The 10-year option is a cost-sharing agreement, not an easement.

Note: Total includes 189,000 acres enrolled in 2004. These acres were in somewhat smaller contracts (188-acre average) with a somewhat larger share in permanent easement (82 percent).

Source: NRCS/USDA.

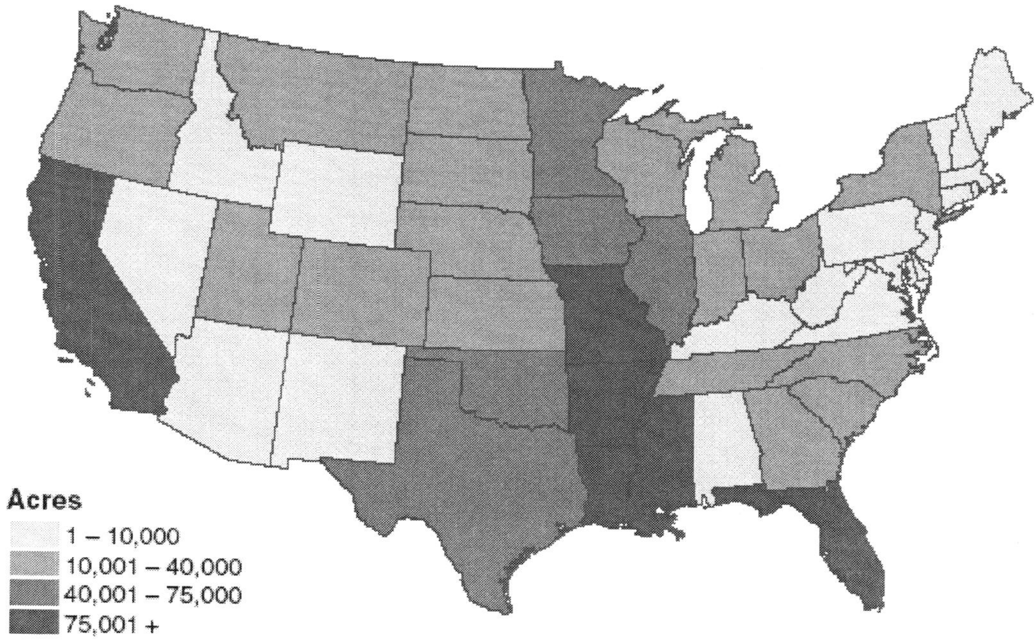

Source: ERS, based on data from the Natural Resources Conservation Service, USDA.

Figure 73. Distribution of WRP land, November 2003.

Table 40. Examples of impacts of the CRP and WRP

Impact	Findings	Sources
CRP: Soil erosion	Soil erosion would increase by 220 million tons/year (60% wind, 40% water) if the CRP were terminated.	Hansen and Barbarika, 2004
CRP: Bird populations	From 1991 to 1995, in 6 Midwest States (IN, KS, MO, MI, NE, IA), bird abundance was 1.4 to 10.5 times greater in CRP land than within row-crop fields.	Best et al., 1997
CRP: Pheasant populations	Ringnecked pheasant numbers in Iowa are believed to have increased 30 percent during the first 5 years of CRP	Riley, 1995
CRP: Duck populations	From 1992 to 1997, the CRP led to an additional 2.4 million ducks in the Prairie Pothole region.	Reynolds et al., 2001
CRP: Monetary measures of value	Improvements in wildlife viewing and pheasant hunting due to the CRP are estimated to be over $700 million per year, plus over $35 million per year from improved water-based recreation.	Feather et al., 1999
WRP: Wildlife and fish	A 7,500-acre project in Oklahoma provides habitat for 256 species, some of which are unusual for the State (such as wood storks and white ibis).	USDA/NRCS
WRP: Flooding	In Missouri, WRP has been used to breech levees on 16,000 acres, which has reduced flood heights and downstream flooding.	
WRP: Threatened and endangered	In Oregon, deep pools were included in a restoration to ensure the survival of the endangered Oregon chub.	

- ***Selecting acres when managing for multiple objectives.*** The CRP seeks to improve more than one environmental resource. Given that more acres are offered to each program than can be accepted, a mechanism that accounts for tradeoffs (between different environmental resources) is necessary. For example, the CRP uses the EBI to choose acres. The weights used in the EBI are based on the informed judgment of a number of scientists and land managers. However, modifications in the EBI—and in what lands are enrolled—could increase the social benefits of the program. For example, Feather et al. (1999) found that using the 15th signup (1997) EBI for all

CRP acres, rather than the simple erodibility criteria used at CRP's inception, increases the value of several outdoor recreation activities by over $350 million per year.

- ***Management modifications.*** As experience with the programs grows, opportunities for fine-tuning emerge. For example, land disturbances (such as grazing and controlled burns) every several years are often necessary to maintain good wildlife habitat (Rodgers and Hoffman, 1997). However, such actions are often costly to the landowner, and require monitoring by USDA. While these concerns have limited the use of such fine-tuning, significant improvements in program performance may be possible with relatively minor changes, such as changes in rental schemes to encourage more active management, or the use of third-party monitoring.
- ***Eligibility expansion.*** The success of voluntary programs such as the CRP and WRP depends on farm participation. For example, the CRP's refocus from erosion to a broader array of conservation priorities increased the pool of eligible acres from about 100 million to 250 million. While this brings in a variety of environmentally valuable lands, it also dilutes the soil conservation emphasis. When many environmental policies and programs exist, this dilution may be positive or negative.
- ***National wetlands goals.*** A current environmental goal is to increase wetland acres nationwide. However, the reduction in Clean Water Act jurisdiction of isolated wetlands underscores the need to use nonregulatory means. The WRP, with its proven record of protecting wetlands, and the CRP's Farmable Wetlands Initiative may acquire additional importance as a means of achieving this national goal.
- ***Upcoming large re-enrollments.*** In 2007 and 2008, over 60 percent (21 million acres) of current CRP contracts will expire. The administrative burden required to replace or re-enroll this acreage could be substantial. In order to reduce these costs while maintaining program flexibility, the USDA plans to use a judicious combination of early re-enrollments and 3- to 5-year extensions.

REFERENCES

[1] Barbarika, Alex. 2001. Conservation Reserve Program: Program Summary and Enrollment Statistics as of August 2001. U.S. Department of Agriculture, Farm Service Agency, Nov. 2001.

[2] Best, L.B., H. Campa, III, K.E. Kemp, R.J. Robel, M.R. Ryan, J.A. Savidge, H.P. Weeks, Jr., and S.R. Winterstein. 1997. Bird abundance and nesting in CRP fields and cropland in the Midwest: a regional approach. Wildlife Society Bulletin 25(4).

[3] Dicks, Michael. 1987. "More Benefits with Fewer Acres Please!," Journal of Soil & Water Conservation, Vol. 42, No. 3, May.

[4] Feather, Peter, Daniel Hellerstein, and Leroy Hansen. 1999. Economic Valuation of Environmental Benefits and the Targeting of Conservation Programs: The Case of the CRP, AER-778, U.S. Dept. Agr., Econ. Res. Serv., April.

[5] Hansen, LeRoy, and Alexander Barbarika. 2004. "The Environmental Benefits of Preserving the CRP." Presented at the Soil and Water Conservation Society annual meeting. Minneapolis, MN, July.

[6] Lubowski, Ruben, Andrew Plantinga, and Roberts Stavins. 2003. "Determinants of Land-Use Change in the United States, 1982-1997." Discussion Paper 03-47, Resources for the Future, Washington, DC.

[7] Osborn, Tim, and Ralph Heimlich. 1994. "Changes Ahead For Conservation Reserve Program," Agricultural Outlook, Vol. 209, July.

[8] Reynolds, R.E., T.L. Shaffer, R.W. Renner, W.E. Newton, and B.D. Batt. 2001. "Impact of the Conservation Reserve Program on duck recruitment in the U.S. prairie pothole region," Journal of Wildlife Management 65(4).

[9] Riley, T.Z. 1995. "Association of Conservation Reserve Program with ringnecked pheasant survey counts in Iowa," Wildlife Society Bulletin 23(3).

[10] Rodgers, Randy, and Richard Hoffman. 1997. Prairie grouse population responses to CRP grasslands: An overview. Draft discussion paper, Kansas Dept. of Wildlife and Parks, Hays, KS.

[11] Roberts, Michael J., and Shawn Bucholtz. 2005. "Slippage in the Conservation Reserve Progam or Spurious Correlation? A Comment," American Journal of Agricultural Economics. 87.

[12] Sullivan, Patrick, Daniel Hellerstein, Leroy Hansen, Robert Johansson, Steven Koenig, Ruben Luboswki, William McBride, David McGranahan, Michael Roberts, Stephen Vogel, and Shawn Bucholtz. 2004. The Conservation Reserve Program: Economic Implications for Rural America, AER-834, U.S. Dept. Agr., Econ. Res. Serv.

[13] Wu, Junjie. 2000. "Slippage Effects of the Conservation Reserve Program," American Journal of Agricultural Economics. 82.

In: Agricultural Resources and Environmental Indicators ISBN 978-1-60021-467-7
Editor: Keith Wiebe and Noel Gollehon, pp. 213-221

Chapter 23

COMPLIANCE PROVISIONS FOR SOIL AND WETLAND CONSERVATION

Roger Claassen

Compliance provisions require Federal farm program participants to conserve soil (on highly erodible cropland) and wetlands. Conservation compliance, which requires application of a soil conservation system on highly erodible cropland, may have reduced annual soil erosion by as much as 295 million tons, accounting for 25 percent of all erosion reduction between 1982 and 1997.

INTRODUCTION

To improve consistency between commodity and conservation programs, compliance provisions require farmers to meet some minimum standard of environmental protection on environmentally sensitive land as a condition of eligibility for many Federal farm program benefits—including farm commodity program payments. Under current compliance requirements, farm program eligibility could be denied to producers who:

- Fail to implement and maintain a Natural Resources Conservation Service (NRCS)-approved soil conservation system on highly erodible land (HEL) that is currently in crop production and was cropped before 1985—a provision known as conservation compliance;
- Convert HEL to crop production without applying an approved soil conservation system—referred to as sodbuster; or,
- Produce an agricultural commodity on a wetland converted after December 23, 1985, or convert a wetland after November 28, 1990, in a way that makes the production of an agricultural commodity possible—referred to as swampbuster.

Producers who violate compliance requirements risk losing all Federal farm programs payments—not just those payments that were (or might have been) made on the HEL or wetland in question.

Sodbuster and swampbuster provisions became effective on December 23, 1985, when the Food Security Act became law. Conservation compliance was implemented over a period of years. By 1990, producers growing crops on HEL were required to have an approved conservation plan. Plans were developed site by site to account for the broad diversity of resource conditions, cropping patterns, and producer preferences. By 1995, producers were required to be actively applying the conservation systems specified in their conservation plan. All three types of compliance have been continued in subsequent farm bills (1990, 1996, and 2002).

Compliance was originally designed to prevent farm support and conservation programs from working at cross-purposes. In the 1970s, evidence suggested that farm commodity programs encouraged crop production on relatively erosive land, even as conservation programs attempted to mitigate erosion (Watts et al., 1983; Reichelderfer, 1985; Heimlich, 1986). Compliance eliminated the farm program incentive to expand production onto HEL.

Compliance mechanisms can also leverage farm program payments for environmental gain—without additional payments—to the extent that producers adopt conservation practices to retain farm program eligibility. Compliance mechanisms are a unique policy tool, distinct from—and in some ways more effective than—conservation payment incentives (e.g., cost sharing). In particular, compliance may be more effective than payments in deterring environmentally harmful actions. For example, a hypothetical subsidy program designed to prevent wetland drainage would require policymakers to pay for protection of all wetlands on agricultural land—a potentially expensive proposition—or decide which wetlands are sufficiently vulnerable to agricultural conversion as to warrant protection—a potentially difficult task (Heimlich and Claassen, 1998b). In contrast, swampbuster penalties are assessed only when a violation occurs, eliminating the need for broad-based subsidies or the need to anticipate the potential for a violation to occur on any given wetland. No direct costs are imposed on producers who comply, although there may be an opportunity cost associated with production forgone on wetlands.

The Compliance Incentive: Producers Weigh Benefits Against Costs

In making decisions about land use and production practices, agricultural producers respond to a range of market signals in the context of available technology, the resources they control (e.g., land), and their own skills and preferences. Any change in land use, investment (e.g., new machinery), or production practices (e.g., reduced tillage) involves both benefits and costs. Likewise, producers who decide to meet compliance requirements are likely to do so because the benefits of compliance outweigh the costs.

Farm Program Benefits and Compliance

Farm program benefits subject to compliance—including farm commodity, disaster, and conservation programs—ranged from $8 billion to $27 billion between 1997 and 2004

(Figure 74). Farmers may also become ineligible for loan and loan guarantee programs that offer reduced interest rates or improved access to credit. Whether these benefits are large enough to leverage conservation depends on whether they exceed the cost of required conservation actions. Because farm program payment levels are set independent of the compliance requirement, there is no guarantee that they will exceed compliance costs. Correlation between payments and conservation needs is critical to the (environmental) success of any compliance requirement, as on highly erodible cropland (Figure 75).

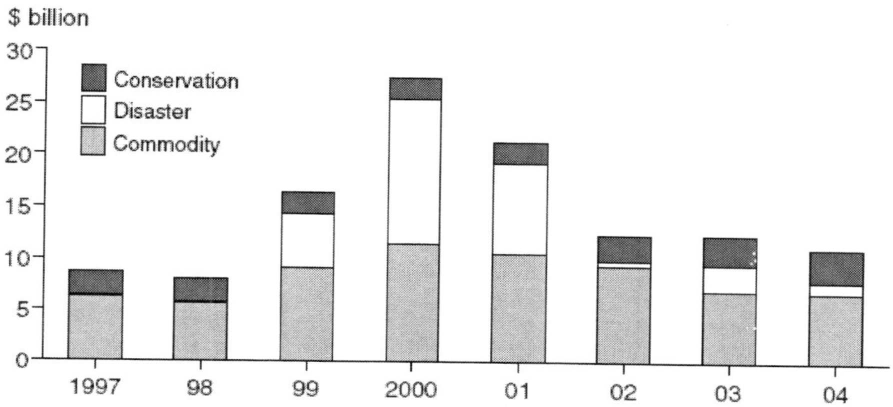

Source: ERS, based on data from the Office of Budget and Program Analysis, USDA, the Highly Erodible Land and Wetland Conservation final rule (7 CFR 12, 61 FR 47019), and communications with national program staff, Farm Service Agency, USDA.

Figure 74. Farm program payments subject to compliance.

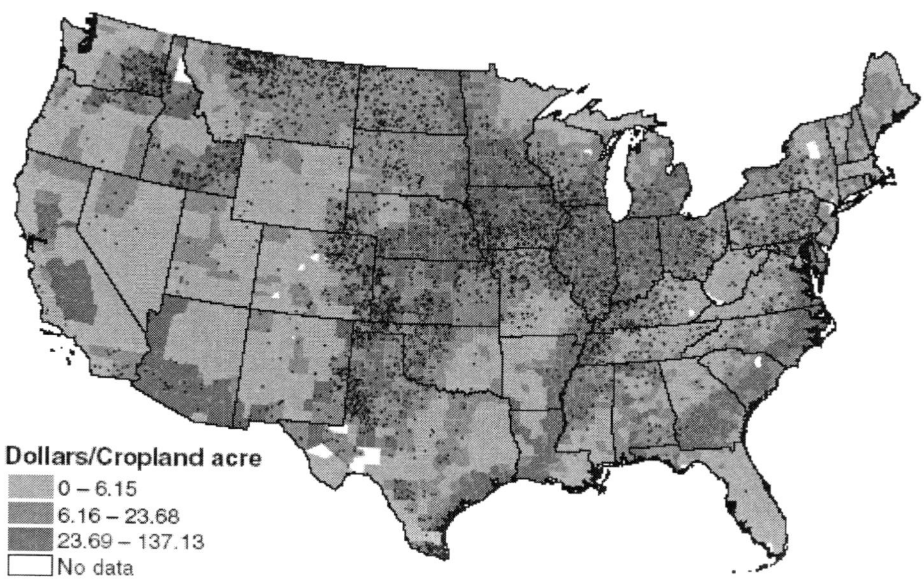

• 1 dot = 25,000 acres of HEL cropland.
Source: Farm Service Agency and NRI.

Figure 75. Commodity payments and Highly Erodible Land (HEL) cropland.

Compliance Costs

Producers may incur direct costs and/or opportunity costs in meeting compliance requirements. Direct costs include the cost of applying and maintaining a conservation system, which depends on the erosion standard to be met and the characteristics of the land (e.g., inherent erodibility). As originally envisioned, conservation systems would be designed to reduce soil erosion to the soil loss tolerance ("T") level—the level that a soil can sustain without long-term productivity damage. Before conservation compliance was implemented, however, USDA determined that reducing erosion to T would be very costly on some land—so costly that a considerable amount of HEL cropland would be unprofitable to farm (Canning, 1994). In the meantime, doubts about the scientific validity of T were being voiced (Cook, 1982) and research showed, increasingly, that water quality damage from sedimentation (which is unrelated to T) exceeded the value of productivity loss (see Ribaudo, 1986; Ribaudo et al., 1990).

As eventually implemented, producers could meet compliance requirements by designing conservation systems to obtain "significant" erosion reduction using "technically and economically feasible" practices, rather than reducing erosion to T. In most cases, conservation systems could be based on inexpensive management practices such as conservation cropping, crop residue management, and conservation tillage. More than half of the HEL cropland acres that meet the Conservation Compliance requirement have approved conservation systems made up of these three practices alone or in combination (USDA, 1999).

Producers may also incur opportunity costs when they refrain from converting additional HEL or wetland that could have been profitably cropped. With wetlands, the opportunity cost equals the value of the land for crop production, less the cost of drainage and land use conversion (e.g., removing trees). HEL not previously cropped can be converted to crop production if an approved conservation system is applied. Compliance cost equals the lower of (1) the opportunity cost of forgoing agricultural production, or (2) the cost of applying a conservation system. On land not cropped before 1985, however, conservation systems must reduce erosion to the T level—a potentially expensive task.

Enforcement

USDA's major enforcement tool is the annual Compliance Status Review (CSR). Each year, through the CSR, USDA field staff assess HEL and wetland compliance on a sample of "tracts" that are identified as part of farms receiving Federal farm program payments subject to compliance provisions. Some tracts are selected at random from the national Farm Service Agency (FSA) database, while others are added by State FSA offices because of potential for noncompliance. For example, tracts on which temporary variances or waivers were previously granted must be checked to establish a return to full compliance.

According to the CSR, overall compliance is high. Based on 1997 CSR data, 95.9 percent of producers subject to compliance were actively applying approved conservation systems. In more recent years, the CSR has shown compliance rates of roughly 98 percent. However, a recent GAO report (2003) identified a variety of deficiencies in the CSR, among them the methods used to select the sample for review, consistency, and clarity of guidance provided to local offices, data handling and analysis, failure to cite producers for significant deficiencies, and inadequate justification for waiver of penalties. This suggests that the actual level of compliance—and whether environmental gains have been realized—cannot be clearly understood using CSR data alone.

What Have Compliance Mechanisms Accomplished?

The rate of soil erosion on U.S. cropland and the rate of wetland drainage for agricultural production have dropped significantly in recent decades. Cropland erosion fell from 3.1 billion tons in 1982 to about 1.9 billion tons in 1997, a reduction of 1.2 billion tons or just under 40 percent. Wind erosion declined by 542 million tons per year (40 percent), while water erosion declined by 633 million tons per year (38 percent). The rate of wetland conversion for agriculture has also declined from 235,000 acres per year during 1974-84 (Dahl and Johnson, 1991) to 26,000 acres per year for 1992-97 (USDA-NRCS, 2002) (see Chapter 6, "Wetlands: Status and Trends").

Although these trends coincide with implementation of compliance mechanisms, the trends alone are insufficient to show compliance's efficacy. Environmental gain can be attributed to compliance mechanisms (or any agri-environmental program) only to the extent that the incentive prompted a change in producer behavior. In other words, we can attribute wetland conservation or erosion reduction to compliance only if the producer or landowner would have done otherwise in the absence of compliance. Because producers respond to a wide range of market and policy incentives, isolating the effect of compliance mechanisms can be difficult. To disentangle these effects, a careful analysis is required (Claassen et al, 2004).

Has Conservation Compliance Reduced Soil Erosion?

Between 1982 and 1997, annual soil erosion from cropland dropped by 1.2 billion tons. (Erosion reduction data are from the National Resources Inventory (NRI)). Of this total, 442 million tons occurred on non-HEL land—where conservation compliance did not apply—leaving 732 million tons (Figure 76). Because compliance was formulated to avoid forcing land out of production, erosion reduction due to land use change (365 million tons, including CRP enrollment) was probably not caused by compliance, leaving 367 million tons. Erosion reduction to levels below the soil loss tolerance (T) level (36 million tons) also cannot be attributed to conservation compliance because conservation compliance required—at most—that erosion be reduced to T. Finally, erosion reduction on farms that do not receive government payments (36 million tons) cannot be attributed to compliance, leaving 295 million tons, or 25 percent of the 1.2-billion-ton reduction in cropland soil erosion between 1982 and 1997.

Furthermore, some erosion reduction may have occurred even in the absence of a compliance requirement. For example, conservation tillage can preserve soil moisture where rainfall is limited and can also reduce machinery, fuel, and labor costs, making it profitable for some producers regardless of its effect on soil erosion. Tillage and planting machinery needed to practice conservation tillage became widely available only in the mid- to late

1970s. Because widespread adoption of new practices often occurs over a long period of time, producers who included conservation tillage in compliance plans may have eventually adopted the practice for economic reasons even without the compliance requirement.

Still, evidence suggests that compliance did have an effect. Reductions in excess erosion (i.e., erosion in excess of T) were larger on farms that received farm program payments than on farms that did not. Excess wind erosion declined by 31 percent on farms receiving payments, but only 14 percent on farms not receiving payments (Figure 77). Excess water erosion dropped by 47 percent on farms receiving payments and by 41 percent on farms not receiving payments.

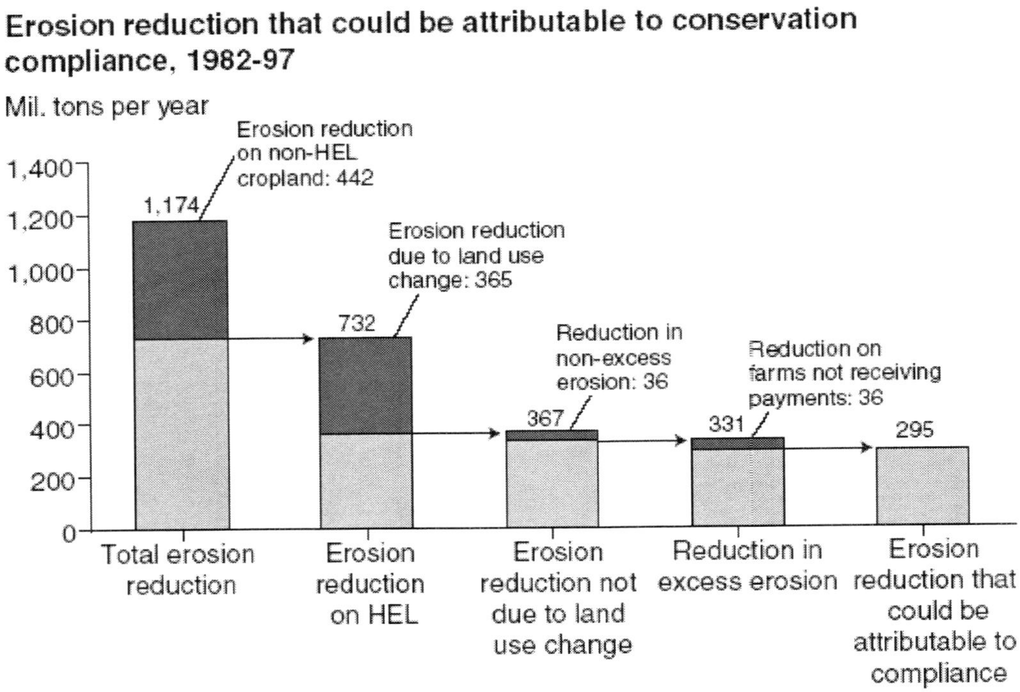

Source: ERS analysis of 1997 NRI and ARMS data.

Figure 76. Erosion reduction that could be attributable to conservation compliance, 1982-97.

Has Swampbuster Slowed Agricultural Wetland Conversions?

Though wetland conversion for agricultural production has declined over time (see part 2), the role of swampbuster is not entirely clear. Swampbuster penalties constrain wetland conversion only when: (1) wetlands are located on farms that participate in Federal programs subject to swampbuster, (2) those wetlands could be profitably converted to crop production in the absence of swampbuster, and (3) other policies (e.g., Section 404 of the Clean Water Act) are not applicable or not effective in deterring wetland conversion.

Many wetlands, ostensibly subject to swampbuster, are in remote areas unlikely to be converted to cropland because they cannot be easily incorporated into an existing farm. Of roughly 90 million acres subject to swampbuster, only 12.9 million are adjacent to existing cropland. These wetlands appear to be located in areas that receive large government payments (Figure 78).

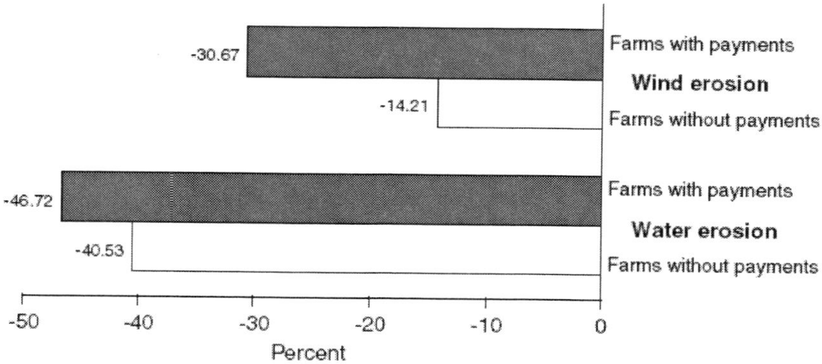

Source: ERS analysis of 1997 NRI and ARMS data.

Figure 77. Change in excess erosion on HEL cropland on farms with and without payments, 1982-97.

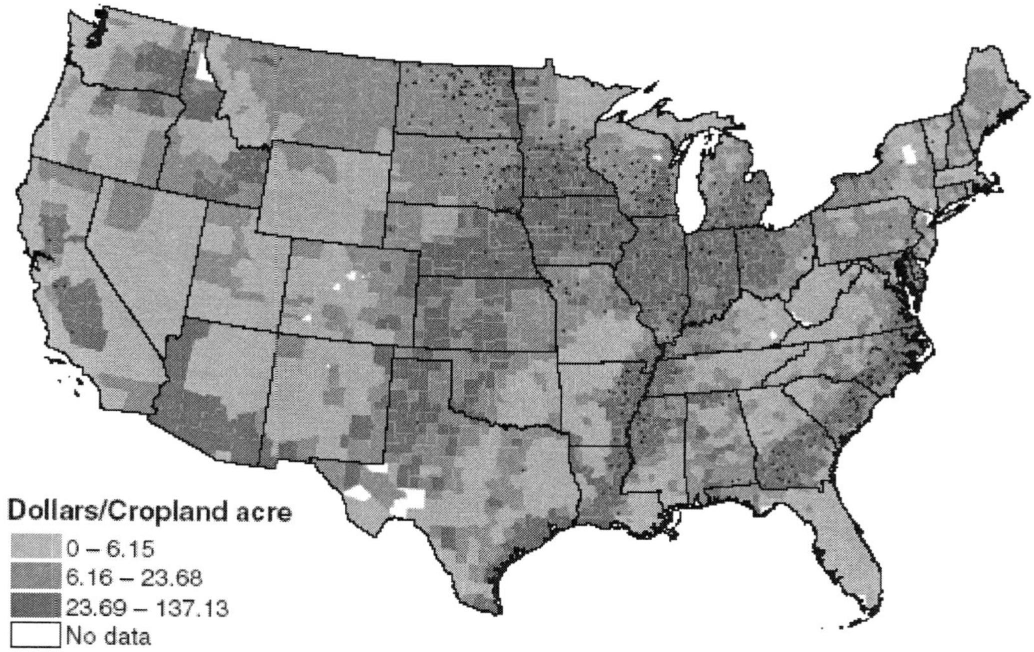

• 1 dot = 25,000 acres of wetlands adjacent to cropland.
Source: Farm Service Agency, USDA.

Figure 78. Commodity payments and wetlands adjacent to existing cropland.

Even so, swampbuster deters conversion only if conversion would otherwise be profitable. On this question, the evidence is mixed. Some researchers have questioned whether wetland conversion for crop production is profitable even without swampbuster (Tolman, 1997; Kramer and Shabman, 1993). Others, using more detailed data on the potential productivity of wetland soils, suggest that there are wetlands that could be profitably converted to crops in the absence of policy constraints. In the absence of swampbuster sanctions, Claassen et al. (2000) estimate that between 1.5 million and 3.3 million acres of wetlands could be profitably converted to crop production under favorable market conditions.

Finally, swampbuster is just one of a number of policies designed to deter or discourage wetland drainage (see Chapter 27 "Federal Laws Protecting Environmental Quality"). Section 404 of the Clean Water Act (CWA) gives the Environmental Protection Agency and the Army Corps of Engineers authority to regulate wetland drainage. Since the January 2001 Supreme Court decision in Solid Waste Agency of Northern Cook County (SWANCC) v. United States Army Corps of Engineers, however, the extent of that authority with respect to isolated wetlands (which are likely to occur in agricultural areas) has been in doubt (Kusler, 2004). While many State and local governments also have wetland laws and regulations on the books and some have increased wetland regulation since the SWANCC decision, many heavily agricultural States have little wetland regulation (Petrie et al., 2001). In these States, swampbuster may be the only remaining policy disincentive to wetland drainage.

Future of Compliance

Compliance mechanisms have seemingly been effective in promoting soil and wetland conservation. While USDA's Compliance Status Review appears to have flaws, these flaws do not mean that compliance rates are low. Evidence from other sources, primarily the National Resource Inventory (NRI), shows that soil erosion on HEL cropland and wetland conversion for agriculture have been sharply reduced. Farms that receive government payments appear to have reduced erosion more sharply than those that do not receive payments, especially in the case of wind-erodible soils. Nonetheless, enforcement of compliance requirements will continue to be a challenge.

Finally, other problems could also be addressed using compliance mechanisms. Claassen et al. (2004) show that a compliance mechanism could be used to address nutrient runoff from land in crop production by encouraging the use of nutrient management or buffer practices. More generally, 86 percent of U.S. cropland is on farms that receive Federal program payments subject to compliance requirements. Thus, compliance could provide leverage in addressing any agri-environmental issue that occurs largely on land in crop production. However, adding multiple or costly compliance requirements could threaten the goal of income support by increasing the cost of farm program participation relative to its benefits.

References

[1] Canning, P. (1994). "Conservation Compliance and Sodbuster," *Agricultural Resources and Environmental Indicators.* U.S. Dept. Agr., Econ. Res. Serv., AH-705, Dec.

[2] Claassen, R., L. Hansen, M. Peters, V. Breneman, M. Weinberg, and others (2001). *Agri-Environmental Policy at a Crossroads: Guideposts on a Changing Landscape.* U.S. Dept. Agr., Econ. Res. Serv., AER-794, Jan.

[3] Claassen, R., P. Johnston, and M. Peters (2000). "Compliance Provisions for Soil and Wetland Conservation," *Agricultural Resources and Environmental Indicators.* U.S. Dept. Agr., Econ. Res. Serv., AH-712, July.

[4] Claassen, R., V. Breneman, S. Bucholtz, A. Cattaneo, R. Johansson, M. Morehart (2004). *Environmental Compliance in U.S. Agriculture: Past Performance and Future Potential.* U.S. Dept. Agr., Econ Res. Serv., AER-832, June.

[5] Cook, K. (1982). "Soil Loss: A Question of Values," *Journal of Soil and Water Conservation.* 37(March-April): 89-92.

[6] Dahl, T.E., and C.E. Johnson (1991). *Status and Trends of Wetlands in the Conterminous United States, Mid-1970's to Mid-1980's.* U.S. Department of the Interior, Fish and Wildlife Service.

[7] Feather, P., D. Hellerstein, and L. Hansen (1999). *Economic Valuation of Environmental Benefits and the Targeting of Conservation Programs: The Case of the CRP.* U.S. Dept. Agr., Econ. Res. Serv., AER-778, April.

[8] Heimlich, R.E. (1986). "Agricultural Programs and Cropland Conversion, 1975-81," *Land Economics.* 62(May): 174-181.

[9] Heimlich, R., K. Wiebe, R. Claassen, R. House, and D. Gadsby (1998). *Wetlands and Agriculture: Private Interests and Public Benefits.* U.S. Dept. Agr., Econ. Res. Serv., AER-765, Sept.

[10] Heimlich, R.E., and R. Claassen (1998a). "Agricultural Conservation Policy at a Crossroads," *Agricultural and Resource Economic Review.* 27(April):95-107.

[11] Heimlich, R.E., and R. Claassen (1998b). "Paying for Wetlands: Benefits, Bribes, and Taxes," *National Wetlands Newsletter.* 20(Nov.-Dec.): 1-15.

[12] Kramer, Randall A., and Leonard Shabman (1993). "The Effects of Agricultural and Tax Policy Reform on the Economic Return to Wetland Drainage in the Mississippi Delta Region," *Land Economics.* 69(3):249-62.

[13] Kusler, Jon (2004). *The SWANCC Decision: State Regulation of Wetlands to Fill the Gaps.* Association of State Wetland Managers, March.

[14] Petrie, M., J-P. Rochon, G. Tori, R. Pedersen, and T. Moorman (2001). *The SWANCC Decision: Implications for Wetlands and Waterfowl.* Ducks Unlimited, Sept.

[15] Reichelderfer, K.H. (1985). *Do USDA Program Participants Contribute to Soil Erosion?* U.S. Dept. Agr., Econ. Res. Serv., AER-532, April.

[16] Ribaudo, M.O (1986). *Reducing Soil Erosion: Off-site Benefits.* U.S. Dept.Agr., Econ. Res. Serv., AER-561.

[17] Ribaudo, M.O., D. Colacicco, L. Langner, S. Piper, and G. Schaible (1990). *Natural Resources and Natural Resource Users Benefit from the Conservation Reserve Program.* U.S. Dept. Agr., Econ. Res. Serv., AER-627.

[18] Tolman, J. (1997). "How We Achieved No Net Loss," *National Wetlands Newsletter.* 19(4):1-22.

[19] U.S. Department of Agriculture, Natural Resources Conservation Service (2002). *National Resources Inventory.*

In: Agricultural Resources and Environmental Indicators ISBN 978-1-60021-467-7
Editor: Keith Wiebe and Noel Gollehon, pp. 223-230

Chapter 24

WORKING-LAND CONSERVATION PROGRAMS

Robert Johansson

Many resource concerns are influenced by agricultural production, and "one-size-fits-all" solutions are unlikely to be effective in addressing them. In many instances, environmental problems like pesticide and nutrient runoff are best addressed on actively cropped lands. More flexible conservation programs for land remaining in production are growing, complementing traditional conservation efforts. Such working-land programs may achieve environmental benefits at relatively low cost, enhancing the amount of environmental gain per conservation dollar spent.

INTRODUCTION

The many and varied resource concerns influenced by agricultural production are often the result of small contributions from many farms over vast areas, and "one-size-fits-all" solutions are unlikely to be effective in addressing them. Policymakers have a wide range of policy instruments to address resource concerns (see Chapter 21, "Conservation Policy Overview"). One tool, land retirement (see Part 5), is and will continue to be an important part of U.S. conservation policy, yet many resource concerns—such as nutrient and pesticide runoff (see AREI Parts 2 and 4)—can be more cost-effectively addressed on the 850 million acres of working cropland and grazing land.

Programs directed at working-land conservation are growing. Much of the 80-percent increase in conservation funding outlined by the Farm Security and Rural Investment (FSRI) Act of 2002 goes toward conservation efforts under two programs that pay farmers for conservation efforts on working lands—the Environmental Quality Incentives Program and the Conservation Security Program.

The Environmental Quality Incentives Program—EQIP

EQIP was established under the 1996 Federal Agriculture Improvement and Reform (FAIR) Act. EQIP's principal objective is to provide producers with assistance that promotes production and environmental quality as compatible goals, optimizes environmental benefits, and helps farmers and ranchers meet Federal, State, and local regulatory requirements.

EQIP provides producers with technical and financial assistance for implementing and managing a wide range of conservation practices for crop and livestock production. Sixty percent of overall EQIP funding is targeted to natural resource concerns related to poultry and livestock production. The remainder is directed toward practices that address conservation priorities on working cropland. Initial funding from 1997 to 2001 was roughly $200 million annually. However, funding for EQIP increased substantially under the FSRI Act—$5.8 billion over 6 years (2002-2007), with annual funding levels increasing from $400 million in 2002 to $1.3 billion in 2007.

Farmers seeking to participate in EQIP complete an application indicating which land will be enrolled, which resource concerns will be addressed, and what practices will be used. Each State or local Natural Resources Conservation Service (NRCS) office ranks applications based on the treatment of priority natural resource concerns; treatment of multiple resource concerns; use of conservation practices that provide long-term environmental enhancements; compliance with Federal, State, local, or tribal regulatory requirements; and the relative cost-effectiveness of the proposed conservation practice. Applications receiving the highest environmental benefit scores based on the ranking criteria are approved for funding.

EQIP uses two types of financial assistance to encourage implementation and management of conservation practices:cost-share and incentive payments, limited to $450,000 per person or entity over a 5-year period. Cost-sharing applies to structural and vegetative practices and may pay up to 75 percent of installation costs, although a 50-percent cost-share is more typical. Examples of eligible practices are grassed waterways, filter strips, waste storage facilities, and caps for abandoned wells. Incentive payments encourage producers to adopt land management practices they may not have otherwise used. Incentive payments are not directly linked to producers'costs; rather, a payment amount sufficient to encourage practice adoption is estimated for each county. Eligible practices include nutrient management, integrated pest management, irrigation water management, and wildlife habitat management.

Distribution of EQIP Funds Geographically

Approximately $2.5 billion has been allocated under EQIP from its inception (FY 1997) through the end of FY 2004. Fund allocation by ERS Farm Resource Region expresses the geographic variation in terms of the natural resource base, products produced, and financial performance (Figure 79).

Although resource concerns vary regionally, payments appear to be distributed among resource regions in rough proportion to the number of farms and value of agricultural production in each region.

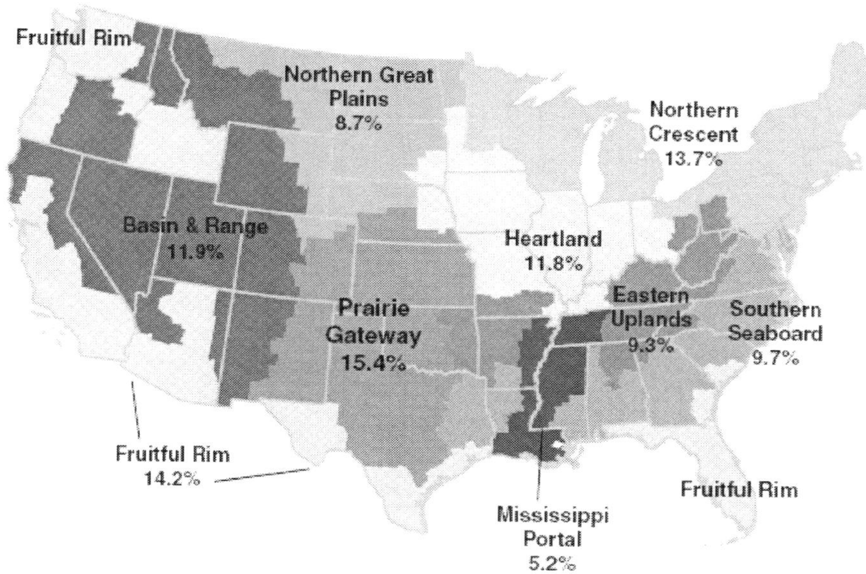

Source: USDA, Economic Research Service, based on FSA data, FY 1997-2004.

Figure 79. Regional distribution of EQIP funds.

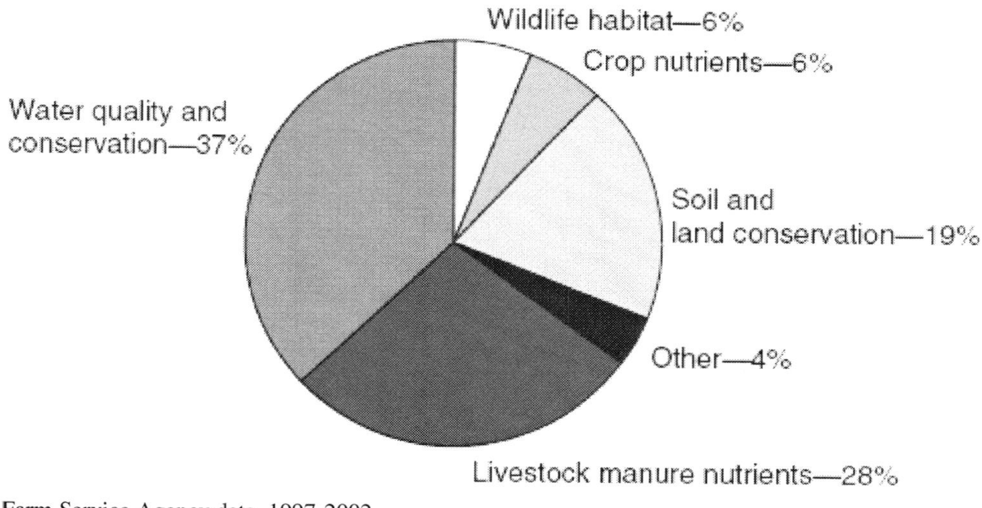

Source: Farm Service Agency data, 1997-2002.

Figure 80. Distribution of EQIP funds by environmental concern, 1997-2002.

Distribution of EQIP Funds by Environmental Concern

Between 1997 and 2002,73 percent of EQIP funds were allocated to geographically defined priority areas: watersheds, regions, or areas of special environmental sensitivity that have significant soil, water, or related natural resource concerns. This regional targeting allowed flexibility in addressing a broad set of environmental priorities (Figure 80), subject to limited funding. At the national level, over one-third of EQIP funds involved water-related conser vation practices, ranging from more efficient irrigation systems to livestock drinking systems. Livestock nutrient management practices accounted for 28 percent of funding, followed by soil erosion and land management with 19 percent of funds. The remaining 16 percent was used to address wildlife habitat management, crop nutrient management, and other concerns.

Regionally, EQIP activity from 1997 to 2004 reflected the confluence of regionally important resource concerns and EQIP priorities. For example, livestock waste management practices obtained the lion's share of EQIP funds in the Northern Crescent, Eastern Uplands, and Southern Seaboard regions, where phosphorus and nitrogen from livestock production (see AREI Parts 2 and 4) far exceed cropland's ability to assimilate these nutrients. However, the presence of excess nutrients does not always result in EQIP funding for livestock manure management. In the Prairie Gateway, which generates substantial manure nutrients on confined animal operations, only 11 percent of EQIP funds were spent on livestock waste management. In the Western States, where water has long been a concern, the majority of EQIP funds were allocated to water resource management. The Northern Great Plains, Basin and Range, Fruitful Rim, and Prairie Gateway all had water quality and water conservation as the main component of EQIP expenditures. In the Heartland, Mississippi Portal, Prairie Gateway, and Southern Seaboard, where much land is subject to soil erosion (see Chapter 5, "Water Quality: Impacts of Agriculture"), a considerable share of EQIP funds was used to prevent soil erosion (Figure 81).

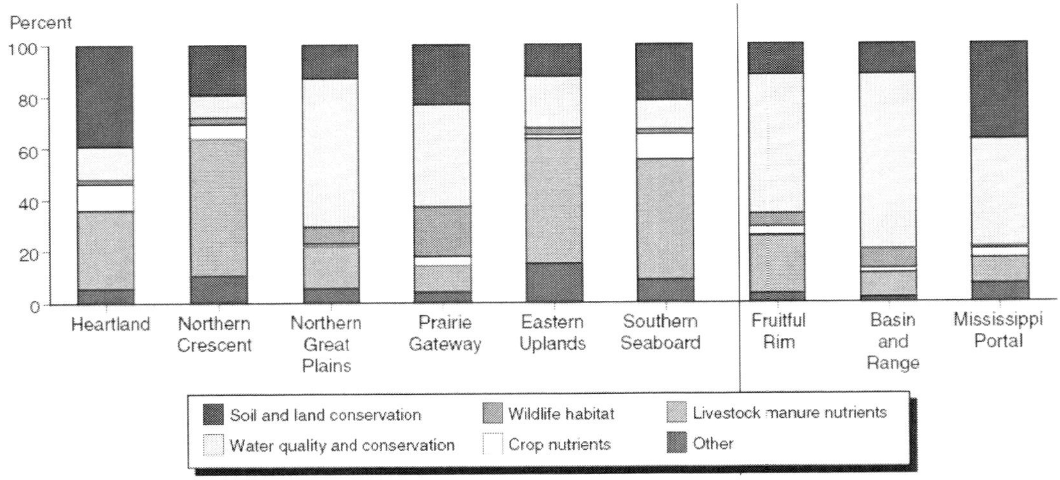

Source: FAS data, 1997-2002.

Figure 81. Distribution of EQIP funds by region and environmental concern.

After 2002, national environmental priorities replaced geographically defined priority areas as a means to screen producers'EQIP applications. These environmental priorities include:

- Reductionsof nonpoint-source pollution, such as nutrients, sediment, pesticides, or excess salinity in impaired watersheds (see Part 2), as well as the reduction of groundwater contamination and the conservation of ground- and surface-water resources (see Chapter 4, "Irrigation Resources and Water Costs");

- Reduction of particulate matter, nitrogen oxides, volatile organic compounds, and ozone precursors and depleters that contribute to air quality impairment;
- Reduction in soil erosion and sedimentation from unacceptable levels; and
- Promotion of habitat conservation for species at risk.

The Conservation Security Program—CSP

CSP was introduced under the 2002 FSRI Act, and the program began in 2004 with a budget of $41 million. CSP addresses familiar conservation issues, but departs from traditional conservation programs in three areas: program eligibility, participation incentives, and selection criteria.

A New Way of Looking at Eligibility

Traditional working-land programs tend toward broad eligibility. EQIP, for example, sponsors adoption of a wide range of practices on many different land types—virtually any type of farm, any type of agricultural land, and any practice found in the NRCS conservation practice handbook can be eligible for funding. Because eligibility has been broad, program decision-makers have used other methods of targeting producers (such as by priority resource concern) or limited participation to stay within budget limits.

In contrast, CSP narrows eligibility to focus on good stewards, and provides payments for the maintenance of some existing conservation practices as well as for the adoption of new practices. Producers become eligible after treating nationally significant resource concerns—soil quality and water quality—using appropriate conservation practices on at least a part of their farm. Depending on the extent to which they have addressed these and other resource concerns, producers may enroll in one of three CSP "tiers."In tier I, producers may enroll only the portion of their farm on which soil and water quality concerns have been addressed by best management practices. Producers who have addressed soil and water quality concerns throughout their farm and agree to address at least one additional resource concern over the life of the contract (5-10 years) are eligible for tier II. Tier III participants must have treated all identified resource concerns—not just soil quality and water quality—with conservation practices before CSP enrollment.

While CSP is a national program, eligibility for any given signup has been limited to specific watersheds. For the initial CSP signup, in July 2004, producers in 18 watersheds were eligible. However, part of the NRCS strategy for CSP implementation is to make every watershed eligible for CSP enrollment. An additional 202 watersheds became eligible for enrollment in 2005, and another 110 watersheds will be eligible for enrollment in 2006 (Figure 82).

Restructuring Participation Incentives

CSP offers several types of payment—some of which are designed to reward past stewardship and assist producers in maintaining previously installed practices. "Stewardship" and "existing practice" payments are based, roughly, on a percentage of the county average rental rate for the specific type of land involved. Practices that are subject to any other maintenance requirement, such as conservation compliance plans, are not eligible for existing practice payments. Implementation of new practices can be cost-shared at a rate of up to 50 percent, 65 percent for limited-resource and beginning farmers. New practices would be required for CSP participants who agree to move to the next higher tier during their CSP contract or for tier II contracts, which require that participants address an additional resource concern over the term of the contract.

[1]Watersheds in Alaska, Hawaii, Puerto Rico, Virgin Islands, Samoa, and Pacific Basin are not shown. For more details see http://www.nrcs.usda.gov/programs.csp.

[2]All 2004 eligible watersheds were eligible in 2005.

Source: USDA, Economic Research Service.

Figure 82. CSP eligible watersheds (2004-2006).

Data from the 2004 CSP signup indicate that two-thirds of CSP payments were for new practices intended to (1) address local resource concerns (e.g., resource concerns other than the nationally significant concerns of soil quality and water quality), and (2) encourage practices or activities that improve or enhance resource quality beyond the minimum (quality criteria) standard. In a number of cases, these payments will be based on environmental performance rather than cost. Environmental indices, such as the soil condition index, will serve as proxies for environmental performance. Payments are to be based on the improvement in index values, ensuring that payments reflect likely environmental gains.

If producer applications exceed available CSP funding, acceptance depends on whether producers meet only the basic requirements of the program (i.e., have addressed soil and water quality concerns) or are willing to implement multiple enhancement practices and activities and move to a higher tier (if not already in tier III).

Table 41. EQIP and CSP designs

Program feature	EQIP	CSP
Budget	2004 contract obligations totaled $718 million. A total of $5.8 billion is authorized for 2002-07.	2004 contract obligations totaled $35.2 million. A total of $6 billion is authorized for 2002-11.
Conservation standard	Producers must address resource concerns to standards in existing NRCS handbook (referred to as "quality criteria").	Standards in existing handbook are a minimum. Through enhancement payments, CSP supports producers in going beyond this minimum standard.
Eligibility	• Both crop and livestock production (in 2003 – 33 percent to crop-related practices; 67 percent to livestock practices). • Emphasis on assisting livestock operations to comply with new Clean Water Act regulations.	• All agricultural land (in 2004 – 67 percent to croplands; 33 percent to range and pasture land). • Animal waste storage or treatment facilities are not eligible.
	• No previous conservation effort required.	• Soil quality and water quality concerns must be addressed before land can be enrolled in CSP.
	• Only practices not started can be funded unless a waiver is obtained at the time of application.	• Existing practices eligible for payments.
	• Available nationally.	• For any given signup, available only in selected watersheds. All 2,119 watersheds to be eligible at least once during 8-year period.
Enrollment screen	Performance-based "offer index." Requests for EQIP finding exceed available budget by 4 to 1.	"Category" system based on level of conservation effort above minimum requirement and performance in terms of soil and water quality criteria.
Participation incentives	Fixed payments: • Cost sharing (typically 50 percent) on structural and vegetative practices; • Incentive payments for management practices. No annual payment limitation, but the sum of all EQIP payments to an individual or entity cannot exceed $450,000.	Fixed payments: • Stewardship and existing practice payment based on rental rates. • Cost-sharing payments for some new practices. Performance-basedpayments. • Enhancements based, in part, on environmental performance. Payments limited by tier: Tier 1 = $20,000 max annual payment; Tier 2 = $35,000 max annual payment; Tier 3 = $45,000 max annual payment.

This new flexibility in conservation program design for working lands and livestock production complements traditional conservation efforts, such as land retirement. In many instances, environmental problems like pesticide and nutrient runoff are best addressed on actively cropped lands. Furthermore, working-land programs may often achieve environmental benefits at a lower cost per acre than under land retirement because land remains in production, thereby minimizing the opportunity cost of environmental gain.

EQIP and CSP–Different Approaches to Similar Concerns

Both EQIP and CSP are designed to address similar resource concerns on working lands. Both of these working-land payment programs are administered by NRCS and in both, payment levels largely determine which eligible producers are willing to participate. Another similarity is that program managers review producers'proposals and decide which ones to accept for program enrollment. This step allows program managers to gather information on potential environmental performance and benefits (and, perhaps, potential to meet other program objectives) and costs directly from farmers –information that can be critical in determining which proposals best contribute to achieving program objectives. However, various program decisions (e.g., budget, eligibility, enrollment screens, and participation incentives) have largely distinguished CSP from EQIP so that now they focus on a wide spectrum of producer types and environmental outcomes (Table 41).

In: Agricultural Resources and Environmental Indicators ISBN 978-1-60021-467-7
Editor: Keith Wiebe and Noel Gollehon, pp. 231-241

Chapter 25

CONSERVATION ON PRIVATE GRAZING LANDS

Marcel Aillery and Dwight Gadsby

Grazing lands provide forage for a significant share of the U.S. animal sector, as well as other economic and ecological services. The preservation and stewardship of private grazing lands has emerged as an important conservation priority, with expanded policy emphasis and funding under USDA farm programs.

INTRODUCTION

USDA has provided technical assistance for grazing systems since the 1930s. An expanded focus on preservation and stewardship of private grazing lands in recent years reflects a growing awareness of their importance to the Nation's environmental health and economic well-being. Achieving USDA conservation objectives for grassland and rangeland resources will involve public/private partnerships in support of sustainable grazing systems.

PRIVATE GRAZING LANDS

Grazing lands are vegetative land area that can be used for the feeding of domestic animals on growing grass, legumes, and other herbaceous plants. Grazing lands encompass a broad range of land types defined by climatic zones, terrain, vegetative cover, and primary land use. Lands used for grazing may include rangelands, grazed forest lands, native grasslands, naturalized and cultivated pasture, and crop and hay lands.

Private grazing land defies easy definition, due to the diversity and multi-use nature of lands used for grazing, distinctions in private ownership and lease arrangements, and land-capability and land-use distinctions across primary sources of grazing land data. Private grazing lands generally include all privately owned, fee-title land used for grazing purposes. Grazed acreage on tribal lands and public lands under State and local jurisdiction, which may be eligible for USDA program assistance, are often subsumed under working definitions of private grazing lands.

EXTENT AND LOCATION OF U.S.GRAZING LANDS

Nearly 35 percent of the total U.S. land area, or 788 million acres of combined Federal and non-Federal lands, was potentially usable for livestock grazing in 1997 (Vesterby and Krupa, 2001). This includes 580 million acres of permanent grassland pasture and rangeland, 68 million acres of cropland pasture, and 140 million acres of forested rangeland (see Chapter 1, "Land Use").

Non-Federal grazing lands—including privately owned land, State and local publicly owned lands, and tribal lands—accounted for 577 million acres in 2002 (Figure 83) (USDA, 2005c). Over 488 million acres of private land were used for grazing purposes in 2002 (Table 42), including pastureland and rangeland (395.3 million acres), forested land used for pasture (31.1 million acres), and cropland (61.8 million). Private grazing lands are located in all States, with heavy concentrations in the Mountain and Plains regions. In the more humid Eastern States, cropland pasture represents a significant share of acreage grazed.

In the West, public lands are used for livestock grazing in designated areas. Federal grazing leases administered by the Bureau of Land Management (U.S. Department of the Interior) and USDA's Forest Service covered 160 million and 95 million acres in 2002 (USDI, 2002; USDA, 2003).

• 1 dot = 25, 000 acres of non-Federal grazing land, which includes pastureland, rangeland, and grazed forest land.

NRI does not collect data for Alaska.

Source: 1997 NRI.

Figure 83. Non-Federal grazing land in the United States, 1997.

SIGNIFICANCE OF GRAZING LANDS FOR THE U.S.ANIMAL SECTOR

Grazing lands provide essential forage for the U.S animal sector. In 1997, roughly 57 million animal-units (AUs)[1] were raised, in part, on forage from grazing lands, accounting for more than 60 percent of AU production on U.S. farms (Table 43). Cow-calf/feeding operations are the dominant grazers, with lesser acreages used for sheep, goats, horses, ponies, mules, burros, donkeys, bison, and llamas.

Table 42. Private grazing lands used by livestock producers, by region, 2002[1]

Region[2]	Pasture, cropland[3]		Pasture woodland		Other pasture and rangeland		Pasture all types
	Million acres	Percent[4]	Million acres	Percent[4]	Million acres	Percent[4]	Million acres
Northeast	1.5	41	0.7	19	1.5	40	3.6
Appalachian	7.2	42	3.4	20	6.3	38	16.8
Southeast	3.6	26	3.6	26	6.5	47	13.8
Lake States	1.9	34	1.5	26	2.2	39	5.6
Corn Belt	7.3	37	3.8	19	8.6	44	19.6
Delta States	3.6	35	2.0	20	4.6	45	10.2
Northern Plains	8.2	10	0.9	1	70.5	89	79.5
Southern Plains	18.5	15	5.8	5	99.1	80	123.5
Mountain	7.2	4	5.7	3	166.7	93	179.6
Pacific	2.8	8	3.7	11	27.7	81	34.3
Alaska/Hawaii	0.0	3	0.0	3	1.6	94	1.7
All U.S.	61.8	13	31.1	6	395.3	81	488.2

[1]Includes farm and ranch operations with $1,000 in annual sales.

[2]Regions are: Northeast (ME, NH, VT, MA, RI, CT, NY, NJ, PA, DE, MD), Appalachian (VA, WV, NC, KY, TN), Southeast (SC, GA, FL, AL), Lake States (MI, WI, MN), Corn Belt (OH, IN, IL, IA, MO), Delta States (MS, AR, LA), Northern Plains (ND, SD, NE, KS), Southern Plains (OK, TX), Mountain (MT, ID, WY, CO, UT, NV, AZ, NM), Pacific (WA, OR, CA), and Alaska/Hawaii (AK, HI).

[3]Reported Census acres of cropland used for pasture were adjusted to reflect the share of animals not raised on farms, as defined by the Census (personal correspondence, Marlow Vesterby, ERS).

[4]Percent indicates the share of each region's grazing land by pasture type.

Source: 2002 Census of Agriculture, NASS, USDA.

An estimated 707,365 animal farms had mostly pastured livestock in 1997, representing 54 percent of all farms with animals (Kellogg, 2002).[2] These farms accounted for $17.2 billion in livestock sales, or 17 percent of U.S. livestock sales in 1997. Most are small operations (less than $10,000 in annual sales) that raise primarily livestock. However, a significant minority raise large numbers of animals; 10 percent of these farms had livestock sales of more than $40,000 (Kellogg, 2002).

---

[1] An animal-unit, defined here as 1,000 pounds of live animal weight, serves as a common unit for aggregating over livestock types.

[2] Farms with mostly pastured livestock were defined to include operations with: (1) fewer than 4 AUs of any combination of animals typically maintained in confined conditions (fattened cattle, milk cows, swine, chickens, and turkeys); (2) 8 or more AUs of cattle other than milk cows and fattened cattle; (3) 10 or more horses, ponies, mules, burrows, or donkeys; or (4) 25 or more sheep, lambs, or goats.

Other farms may also use grazing lands. Farms with few animals—raised primarily for home consumption or local markets—are likely to depend on pasturing for feed needs. Pastured livestock are more common on operations of fewer than 50 AUs (Table 43). Some confined livestock farms (predominantly cattle feedlot and dairy operations) may depend on forage grazing for some animals over part of the year, and may have large numbers of pastured livestock.[3] An increasing concentration of unconfined animals on larger operations (greater than 300 AUs) over 1982-97 (Table 43) mirrors a similar trend in confined animal production (Kellogg, 2002).

Table 43. Number of animal units1, total and unconfined, by operation size, 1982 and 1997

Farms by number of AUs	1982 All animals	Unconfined animals	Percent share unconfined	1997 All animals	Unconfined animals	Percent share unconfined
		Mil. AUs			Mil. AUs	
< 25	7.3	6.7	92	5.4	5.2	96
25 -< 50	9.5	7.5	79	7.3	6.4	87
50 -< 150	29.0	17.5	60	21.5	14.9	69
150 -< 300	17.1	10.3	60	16.0	9.9	62
300 -< 1,000	16.9	10.9	65	20.3	12.1	60
1,000 +	15.8	7.2	46	24.9	8.8	35
Total	95.6	60.1	63	95.3	57.3	60

[1]Animal-unit numbers by farm size were calculated based on beef and dairy cattle, swine, and poultry. Other animal types that are typically pastured—including sheep, goats, horses, ponies, mules, burros, and donkeys—represent an additional 3.5 million AUs.

Source: Adapted from Kellogg et al., 2000, and Kellogg, 2002, based on agricultural census data for 1982 and 1997.

Additional Benefits of Grazing Lands

Grazing lands support other activities in addition to livestock production that contribute to rural economies, such as hunting and fishing, wildlife viewing, and other ranch-based recreation. Fees generated from these uses supplement income for some animal producers and may help sustain operations. Grazing lands are also regarded as an integral part of the cultural heritage and identity of many rural communities.

Grazing lands, where properly managed, provide important ecological functions. Grazing lands help to maintain habitat and migration corridors for wildlife, supporting a rich biodiversity of plant and animal species. As grazing lands account for large acreages in many U.S. river basins, they are important in hydrologic processes involving streamflow, aquifer recharge, and water filtration. In addition, grazing lands sequester substantial amounts of atmospheric carbon. Potential gains from cropland conversion to grassland have been considered in the context of U.S. policy on climate change mitigation (Follett et al., 2001).

[3] USDA estimates assume that confined livestock may be pastured for up to 45 days a year (Kellogg et al., 2000).

Conservation Policy Concerns

Two broad areas of policy concern involve the loss of private grazing land area and resource degradation on grazing lands.

Area Loss

Conversion of grassland for crop production and developed uses has reduced the extent of native grasslands in the U.S. by roughly 50 percent, with significant fragmentation of remaining grassland resources (Conner et al., 2001). Losses have been greatest in the historic savanna and tall-grass prairies of the Midwest and Central Plains, and relatively less in the arid West where nonirrigated cropping potential is limited and much of the land is publicly owned. While the rate of loss has slowed in recent decades, area in grasslands and other grazing land resources continues to decline. From 1982 to 2002, acreage in non-Federal grazing lands fell by 5 percent according to USDA's Natural Resources Conservation Service—from 611.0 to 577.7 million acres—including reductions in pastureland (13.7 million acres), rangeland (10.2 million acres), and grazed forestland (9.5 million acres) (Figure 84).

Cropland expansion has fueled much of the grassland conversion, particularly in years of strong crop demand. More recently, increases in population and income have driven substantial exurban development in grasslands (Conner et al., 2001). Reductions in grazing land resources nationwide, however, may mask variability in land-use coverage over time. In marginal cropping areas, cropland conversions (and reconversion to grassland) may be influenced by relative returns to crop and livestock production and changes in agricultural policies (see Chapter 22, "Land Retirement Programs"). In some locations, Federal cropland retirement initiatives have resulted in increased grassland area, which may be grazed under specified conditions.

Resource Degradation

Of the remaining grassland resources in private ownership, much of this acreage has been degraded due to overgrazing, fire suppression, invasive species, and other factors (Conner et al., 2001). Degradation of the land resource is reflected in reduced forage productivity for livestock and environmental damages, both on and off the site.

Environmental effects of livestock grazing may include excessive foraging and trampling of vegetative cover, streambank erosion, and sediment/ nutrient loadings to water bodies that may harm riparian and upland habitat. Livestock grazing has been cited as a factor in the decline of threatened and endangered species under the Endangered Species Act. Of 663 species identified as affected by agricultural activity (as of September 1995), livestock grazing was a factor in 171 listings (26 percent) (Lewandrowski and Ingram, 2002).

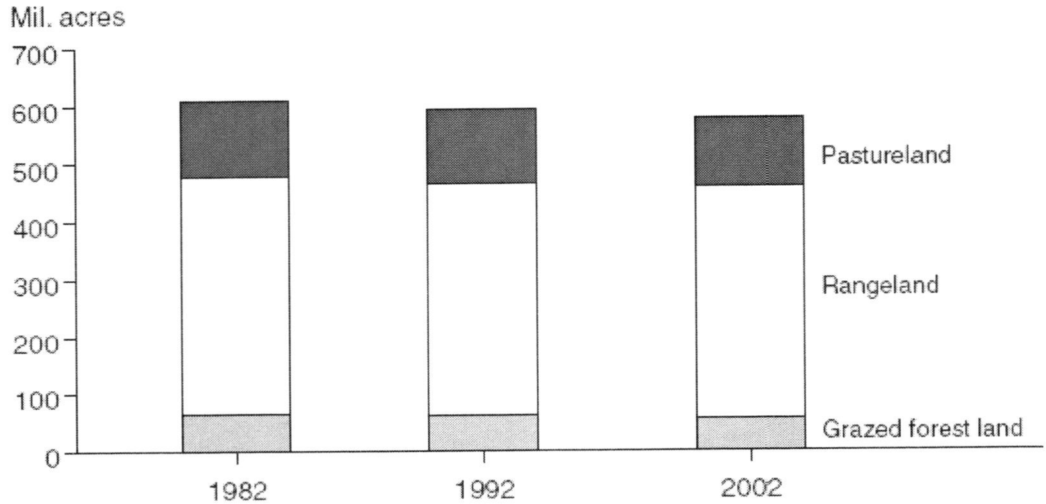

Source: 2002 National Resources Inventory, NRCS (USDA, 2005c).

Figure 84. Trends in non-Federal grazing land, 1982-2002.

Considerable policy attention has focused on animal waste management in recent years, with new Federal regulations enacted in 2003 for the largest confined animal operations (see AREI Parts 2 and 4). Waste from unconfined (pasture-based) operations remains largely unregulated, although it may impair local water quality. Roughly half of the manure nutrients produced on U.S. animal farms was generated by unconfined livestock in 1997 (Figure 85), including 3.3 million tons of manure nitrogen (51 percent) and 1.0 million tons of manure phosphorus (54 percent).[4]

Pathogen contamination from animal waste is an important public health issue. A recent USGS study examined water quality effects of fecal coliform bacteria from confined and unconfined animal operations. While loadings are largest in drainages downstream of confined operations (reflecting the volume of concentrated waste), manure from pastured animals contributes much more fecal coliform bacteria to streams per AU nationwide (Smith et al., 2004).

Improved Grazing Systems

Increased policy attention has focused on livestock grazing systems that are environmentally and economically sustainable. Field studies suggest that grazing lands can be managed to enhance forage productivity while preserving environmental quality (USDA, 2005a; AFGC, 2001). Practices undertaken as part of an improved grazing system include rotational grazing to allow grass rejuvenation; fencing to restrict livestock access in sensitive areas; watering facilities to remove livestock from riparian areas; windbreaks and shelterbelts to disperse herds; manure storage facilities for temporary confinement areas; filter strips to

[4] These estimates likely understate the potential impact of manure nutrients on grazing lands, as (1) a share of animals on confined operations are pastured for a portion of the year; (2) recoverable manure from confined operations may be land-applied on pasture, either on or off the source farm; and (3) values do not reflect manure production from all animal types typically pastured.

intercept runoff from heavy-use areas; improved grass and legume cultivars; improved nutrient management practices; and integrated pest management strategies.

Producer returns may also increase from improved grazing practices. Benefits may include additional quantity and quality of forage; healthier livestock and lower veterinary costs; better monitoring of livestock, resulting in earlier problem detection; higher weaning weights; and reduced problems with noxious weeds and other undesirable plant species. In many cases, however, public incentives will be required to encourage adoption of recommended grazing practices, particularly where benefits primarily occur offsite.

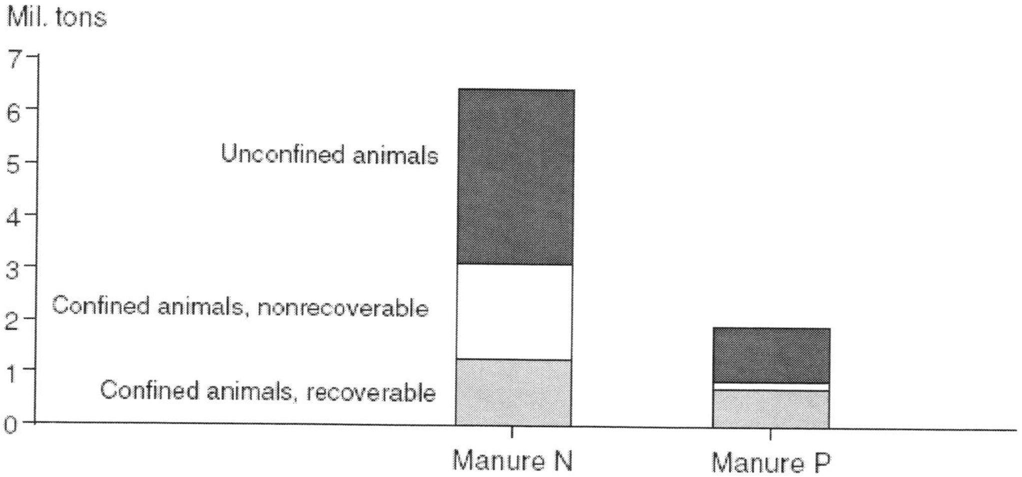

[1]Based on beef and dairy cattle, swine, and poultry.
Source: Adapted from Kellogg et al., 2000.

Figure 85. Manure nitrogen and phosphorus production in the U.S. animal sector1, 1997.

FEDERAL SUPPORT FOR CONSERVATION ON PRIVATE GRAZING LANDS

The Federal Government provides conservation information and technical assistance for private grazing lands, primarily through USDA's Natural Resources Conservation Service (NRCS). Non-Federal grazing lands constitute about half of the total land on which NRCS provides technical assistance (USDA, 1997). According to NRCS, roughly 355 million acres of private grazing lands are in need of some form of conservation treatment (USDA, 2001). NRCS technical assistance is funded primarily through the Conservation Technical Assistance program, which allocated roughly $100 million toward grazing-related initiatives in FY2004 (USDA, 2005f). Development of soil surveys and ecological site descriptions for grazing lands, and approved conservation plans for grazing systems, will likely be emphasized in the coming years (USDA, 2005f).

Comprehensive nutrient management plans (CNMPs), designed to minimize water quality impairment from manure nutrients, are an important element of an overall conservation plan for many animal operations. Of an estimated 257,201 farms with confined animals that are likely to need CNMPs, roughly one-fourth had pastured animals as the dominant type (USDA, 2003).[5] Average annual CNMP costs per farm with pastured livestock were estimated at $1,450 (USDA, 2003).

The Environmental Quality Incentives Program (EQIP), introduced in 1996 and extended under the Farm Security and Rural Investment Act of 2002 (2002 Farm Bill), provides technical and financial assistance to address natural resource concerns on working farms and ranches (see Chapter 24, "Working-Land Conservation Programs"). Cost-share and incentive payments under 5- to 10-year contracts are available for eligible practices in an approved conservation plan. Sixty percent of EQIP funding under the 2002 Farm Bill is targeted to livestock production, with improved grazing systems as an important element. In 2004, more than $95 million in EQIP cost-sharing was approved for practices involving unconfined livestock (USDA, 2005d) (Table 44).

The 2002 Farm Bill includes several other programs that support conservation on grazing lands:

- The Grassland Reserve Program (GRP) targets grazing operations on private grasslands. The GRP, administered jointly by NRCS and the Farm Service Agency (FSA), was authorized by the 2002 Farm Bill. The program is designed to preserve grasslands for livestock grazing and other uses. Enrollment options include permanent and long-term (30-year) easements with a single upfront payment and long-term rental agreements (10, 15, and 30 years) with annual payments. An approved grassland resource management plan is required for all enrolled lands, with compensation for the use of approved practices. Program funding of $254 million is authorized over FY 2002-07, with a total enrollment cap of 2 million acres nationwide.
- The Conservation Reserve Program (CRP), administered by FSA and NRCS since 1985, targets removal of environmentally sensitive lands from agricultural production under 10- to 15-year lease agreements (see Chapter 22, "Land Retirement Programs"). Much of the CRP enrollment involves marginal croplands in grassland areas of the Plains. Enrolled lands are planted to native grasses and other vegetative cover, and pasturing is permitted (subject to reduced CRP payments) as part of an approved conservation plan.
- The Conservation Security Program (CSP), administered by NRCS since 2002, provides financial and technical assistance to farmers and ranchers recognized as exemplary land stewards (see Part 5). Pasture and rangeland accounted for more than 30 percent of total acres approved for contracts in FY 2004 (USDA, 2005e).
- The Farm and Ranch Lands Protection Program (FRPP), administered by NRCS since 1996 (Farmland Protection Program prior to 2002), helps maintain working cropland and grazing lands by providingmatching funds to State, tribal, and local governments, as well as nongovernmental organizations, for conservation easement acquisition (see Chapter 5, "Farmland Protection Programs").

---

[5] CNMPs are required for all concentrated animal feeding operations under EPA regulations, estimated to apply to 15,500 of the largest operations. However, USDA encourages all animal operations to develop CNMPs.

Table 44. EQIP contracts, total expenditures, and cost-share payments for selected practices associated with livestock grazing, 1997-20031

Conservation practice	Number of contracts	Total expenditures for practice	EQIP cost-share Payments
		$ million	
Fencing	48,330	156.5	103.9
Prescribed grazing	38,721	56.9	44.9
Trough or tank	35,646	57.3	38.0
Pasture and hay planting	35,119	88.7	58.0
Brush management	18,849	85.5	51.7
Range planting	5,683	17.0	10.8
Spring development	4,908	9.5	6.5
Windbreak/shelterbelt establishment	3,627	6.0	4.1
Upland wildlife habitat management	1,989	1.7	1.3
Prescribed burning	1,733	2.3	1.7
Animal traits and walkways	1,616	5.7	4.1
Stream crossing	926	2.6	1.7
Riparian forest buffer	769	.9	.7
Animal use area protection	754	3.7	2.2
Grazing land mechanical treatment	443	1.1	.7
Windbreak/shelterbelt renovation	433	.9	.6
Planned grazing system	387	1.0	.7
Pasture and hayland management	330	.5	.3
Stream channel stabilization	164	.9	.6

[1]Based on NRCS conservation practices identified in EQIP contracts for producers reporting animals, 1997-2003.

Source: USDA EQIP database.

The Grazing Lands Conservation Initiative (GLCI) is a nationwide collaboration of stakeholders—farm and ranch organizations, State and Federal entities, tribes, and environmental interests—working to complement conservation programs through research, education, and technical assistance. Program funding is supported by congressional appropriations, with $23.5 million in FY 2004 (USDA, 2005f).

USDA's Agricultural Research Service directs research on sustainable grazing systems through the Rangeland, Pasture, and Forages (RPF) National Program (USDA, 2005a). The RPF program encompasses a broad range of interdisciplinary research projects involving collaboration across Federal and State agencies and land-grant universities.

FACTORS AFFECTING CONSERVATION ADOPTION ON PRIVATE GRAZING LANDS

Returns to ranching in some areas may limit investment in conservation practices, particularly for smaller operations with limited capital.[6] Adoption incentives may be inadequate without increased livestock returns, as when measures are designed to protect habitat. Incentives may also be limited for lands grazed under a lease agreement or informal arrangement, where the operator does not capture long-term benefits (Lewandrowski and Ingram, 1999).

USDA farm programs have historically supported returns to crop producers through price supports and mitigation of crop risk. Farm support payments have largely been decoupled from production since 1996, but certain payments (such as loan deficiency payments) continue to be linked to crop production. Where USDA programs enhance crop returns relative to livestock grazing in marginal cropland areas, program incentives may have the unintended consequence of encouraging grassland conversion to crop production and discouraging reversal to grasslands (Conner et al., 2001). ERS analysis suggests that Federal crop insurance has contributed to crop-land development in marginal cropping areas, although acreage effects have been small (Claassen et al., 2005).

Policy mechanisms for conservation on private grazing lands are largely nonregulatory. While large confined animal operations are regulated as a point-source for waste discharge, onsite environmental effects of grazing are more diffuse and consequently less subject to mandatory controls. Adoption of conservation measures on grazing lands has relied largely on technical assistance and voluntary incentives, without regulatory or compliance mechanisms to ensure environmental standards.

The proliferation of ranchettes (subdivisions of large rural tracts) in many areas represents a further challenge for conservation policy. Conservation concerns can be particularly significant, as smaller land holdings may be overstocked with animals relative to carrying capacity and manure-nutrient uptake. As owners do not generally depend on livestock for income, financial incentives may be less effective in encouraging improved grazing systems. Effective strategies may require coordination of conservation activities across multiple landowners.

Many Western ranches use a mix of Federal, State, and private lands for livestock grazing over the course of a year. Access to public lands is often critical to providing private parcels adequate time to recover within a rotational grazing regime. For much of the West, the success of conservation measures on private grazing lands may be linked to grazing policies for public lands.

REFERENCES

[1] American Forage and Grassland Council (AFGC) (2001). Stewardship for the 21st Century: A Report on America's Forage and Grassland Resources and Needs, March.

[6] Over 1998-2003, average returns above total costs for cow-calf operations in the U.S. were considerably less than returns to wheat production (USDA, 2005b).

[2] Claassen, R., R. Lubowski, and M. Roberts (2005). "Extent, Location, and Characteristics of Land Cropped Due to Insurance Subsidies." Economic Research Service, USDA. Selected paper for the annual meeting of the American Agricultural Economics Association, Providence, RI, July.

[3] Conner, R.A., A. Seidl, L. Van Tassell, and R.N. Wilkins (2001). United StatesGrasslands and Related Resources: An Economic and Biological Trends Assessment.

[4] Kellogg, R.L., C.H. Lander, D.C. Moffitt, and N. Gollehon (2000). Manure Nutrients Relative to the Capacity of Cropland and Pastureland to Assimilate Nutrients: Spatial and Temporal Trends for the United States. Natural Resources Conservation Service and U.S. Dept. Agr., Econ. Res. Serv., Dec.

[5] Kellogg, R.L. (2002). Profile of Farms with Livestock in the United States: A Statistical Summary. Natural Resources Conservation Service, USDA.

[6] Follett, F.F., J.M. Kimble, and R. Lal (editors) (2001). The Potential of Grazing Lands to Sequester Carbon and Mitigate the Greenhouse Effect, Lewis Press, Boca Raton, FL.

[7] Lewandrowski, J., and K. Ingram (1999). "Policy Considerations for Increasing Compatibilities between Agriculture and Wildlife," Natural Resources Journal, Vol. 39, No. 2, pp. 229-267.

[8] Lewandrowski, J., and K. Ingram (2002). "Restricting Grazing on Federal Lands in the West To Protect Threatened and Endangered Species: Ranch and Livestock Sector Impacts," Review of Agricultural Economics, Vol. 24, No. 1, pp. 78-107.

[9] Smith, R.A., R.B. Alexander, G.E. Schwarz, and M.C. Ierardi (2004). "Effects of Structural Changes in U.S. Animal Agriculture on Fecal Coliform Contamination of Streams." U.S. Geological Survey. Presentation before the American Geophysical Union, San Francisco, CA, Dec.

In: Agricultural Resources and Environmental Indicators ISBN 978-1-60021-467-7
Editor: Keith Wiebe and Noel Gollehon, pp. 243-251

Chapter 26

FARMLAND PROTECTION PROGRAMS

Cynthia Nickerson and Charles Barnard

Public support for protection of farmland—cropland, pasture, and range-land—is growing, as private decisions to convert land may not account for rural amenity and other nonmarket benefits of farmland. Protection programs include both regulatory and voluntary measures. Whether the benefits of farmland protection exceed program costs may depend greatly on local conditions.

INTRODUCTION

Expansion of land in urban uses often encroaches on cropland, pasture, and rangeland. When these types of farmland are converted to urban uses, the ability of the land to produce agricultural outputs is lost. Such losses are the focus of growing public financial support for farmland protection. All 50 States have enacted one or more farmland protection programs to help slow the conversion of farmland to developed uses.

If farmland only produced agricultural commodities, the normal workings of the land market would optimally allocate land between farming and urban uses. However, farmland also provides a number of other benefits, or rural amenities, including open space, scenic views, rural agrarian character, and wildlife habitat. These nonmarket benefits are not typically accounted for in the land market, as landowners are seldom able to extract payment from anyone for providing these amenities. Consequently, landowners may not take the social value of these amenities into account when considering whether to develop land for urban-related purposes.

TRENDS IN FARMLAND LOSSES

While farmland converted to urban uses comes from a large base, urban areas have grown rapidly from a small acreage base (see Chapter 1, "Land Use"). On average, 2.2 million acres of farmland per year were converted to urban uses between 1992 and 2001, versus 1.1 million acres per year during the previous decade (Vesterby and Krupa, 2001). Still, this annual rate represents barely 0.2 percent of the Nation's 1.03 billion acres of cropland, grassland, pasture, and rangeland, and suggests little threat to the Nation's capacity to produce food and fiber (Barnard, 2000).

Rapid urban development since World War II has been fueled primarily by population and economic growth, which has occurred in conjunction with increased automobile ownership, declines in average household size, and an increase in average residential lot sizes beyond the urban fringe (Heimlich and Anderson, 2001). The movement of urban populations to suburban locations has also increased development pressures (Barnard, 2004). Despite more than doubling since 1960, urban area made up less than 3 percent of U.S. land area (excluding Alaska) in 1997. Developed area—which includes urban areas plus large lot development, development in rural areas, and rural roads and transportation—made up slightly more than 6 percent in 1997 (Vesterby and Krupa, 2001).

Land moves into and out of different uses for a variety of reasons. Movements of land into urban uses, however, tend to be permanent. Once farmland is developed, it is typically economically infeasible to revert back to farming. In 1982-97, 22.7 million acres of farmland were converted to forest, versus 13.9 million acres converted to urban uses.[1] About 5.4 million acres of land converted to urban uses were prime farmland. However, the share of land converted that was prime (22 percent) was very similar to the share of the land base that was prime in 1982 (20 percent), so prime farmland was not disproportionately converted (Figure 86).

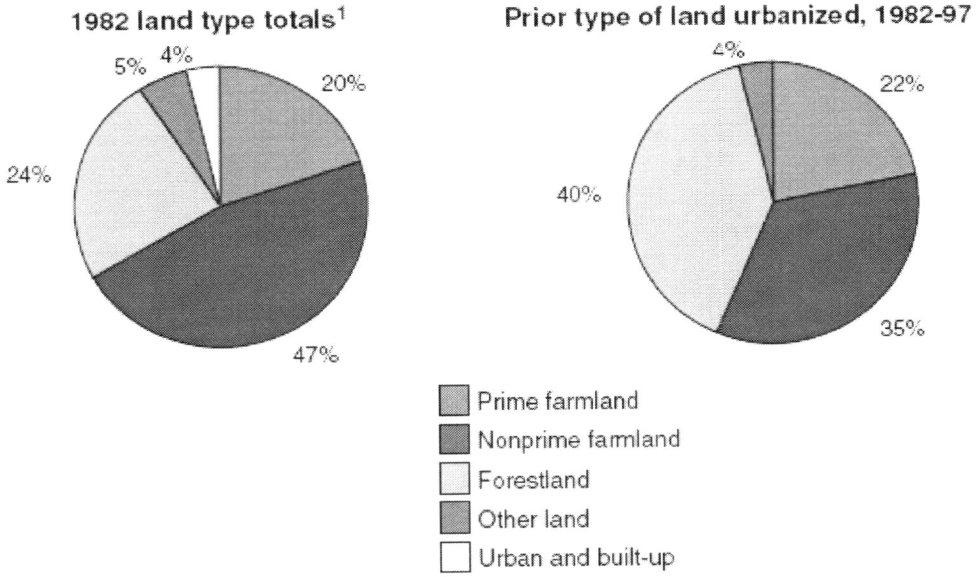

[1]Excludes about 400 million acres in federally owned land and 365 million acres in Alaska.
Source: ERS analysis of National Resources Inventory data, 1982-97.

Figure 86. Land type and composition of change, 1982-97.

[1] Some of the reported shift to forest use is likely due to reclassifications. As trees reach a 10 percent canopy level, they are classified as forest, even though the land may still be used for grazing.

The amount of land in cropland uses remained nearly constant nationwide between 1945 and 1997, at about 20 percent of U.S. land. Yet, some regions have consistently lost cropland (Figure 87). The Northeast lost 46 percent (11.6 million acres) of the cropland that existed in 1945, the Southeast lost 33 percent (9.0 million acres), Appalachia lost 20 percent (7.0 million acres), and the Lake States, 12 percent (5.5 million acres). Western regions, however, added 12 percent (37.1 million acres). Losses in the East are likely due to increased urbanization, while Western gains are due in part to federally subsidized irrigation water (Vesterby and Krupa, 2001).

Losses in grassland pasture and range in 1945-97 exceeded 70 percent (7.3 million acres) in the Northeast. Causes include natural regeneration of forests and losses of grassland to urban development (Vesterby and Krupa, 2001). Grassland losses in the West were 10 percent (61.4 million acres), due largely to nonpermanent conversions to cropland. (See the "Major Land Uses" data product on the ERS website for more information.)

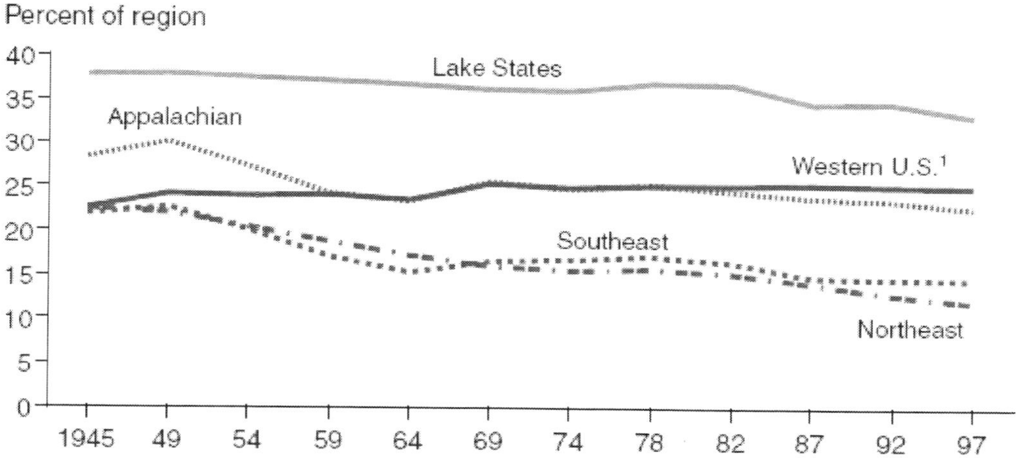

[1]Includes Northern and Southern Plains, Mountain, Pacific, Corn Belt and Delta regions.
Source: Vesterby and Krupa (2001).

Figure 87. Share of region that is cropland, 1945-97.

FARMLAND PROTECTION POLICIES AND TOOLS

Because private land use and conversion decisions may not account for rural amenity and other nonmarket benefits provided by farmland, government agencies and other organizations adopt policies and programs to protect farmland. Land use management is a local prerogative by tradition and law, and every State has enacted measures that help protect farmland. An ERS analysis of the "purpose clauses" of State farmland protection laws and programs found that protecting rural amenities was cited by 36 States, along with protecting local food supplies (30 States), protecting environmental services—including water and air quality (29 States), protecting the local economy's natural resource-related jobs (23 States), and maintaining orderly development (18 States). The focus on protecting rural amenities most often stemmed from goals relating to the protection of open space and rural/agrarian character.

Local jurisdictions and nonprofit organizations have adopted an expanding array of farmland protection programs since World War II. Agricultural/rural residential zoning defines minimum parcel sizes and may include limitations that restrict use to farm-related activities (farm family and labor housing, processing, and marketing). Another regulatory approach is right-to-farm laws, which protect farmers from nuisance lawsuits brought by neighbors objecting to normal farm activities, and sometimes from local government-imposed ordinances that unreasonably restrict agricultural activities

Voluntary approaches include preferential assessment, which allows jurisdictions to assess agricultural land for property tax purposes at its value in current agricultural uses instead of its full market value for potential urban (developed) uses. In some cases, landowners must forgo development for a specified time period. Preferential assessment laws were first enacted at the State level in Maryland in 1956; by 1989, they had been adopted by all 50 States. Other voluntary approaches include agricultural districts, in which enrolled landowners maintain the land in an agricultural use for a specified term, in exchange for property tax relief, insulation from nuisance complaints, and other benefits; Purchase of Development Rights (PDR) programs, in which landowners sell the rights to develop the land; and Transfer of Development Rights (TDR) programs, in which landowners in locally designated "sending areas" privately negotiate to sell development rights to developers who use them to develop at higher densities in locally designated "receiving areas." Use of these incentive-based mechanisms avoids the property rights issues that have hampered regulatory programs.

TRENDS IN FARMLAND PROTECTION

State and local governments spend millions of dollars annually on farmland protection programs. For example, ERS estimated that costs incurred through use value assessment programs (a "tax expenditure") range from about $25,000 annually in Wyoming to $218 million annually in California. The national total is almost $1.1 billion annually (Heimlich and Anderson, 2001).

Another major outlay is State and county PDR programs. Nineteen States have State-level PDR programs, and at least 41 local jurisdictions operate separate programs in 11 States (AFT, 2004a and 2004b). The average easement cost in State PDR programs was about $1,400 per acre, and nearly $2,000 per acre in local PDR programs. However, PDR expenditures are one-time expenditures to restrict development over the long term (or permanently).

The most active State and local PDR programs are in the Northeast. Maryland, Massachusetts, New Jersey, and Pennsylvania account for 76 percent of State-level PDR expenditures to date and 58 percent of the acres preserved to date in State programs (AFT, 2004a). Especially active programs elsewhere are county-level programs in Sonoma County, CA, and King County, WA.

ERS estimates all State PDR programs to average $123 million in spending annually. State PDR programs have cumulatively preserved nearly 1 million acres of farmland at a cost of nearly $1.4 billion since the late 1970s (Figure 88) (AFT, 2004a). This is slightly more than the annual tax receipts that are forgone through use value assessment (when capitalized at 4 percent, the 1995 value of U.S. public expenditures on use value assessment is estimated to be $27 billion). The amount of land preserved represents less than 1 percent of cropland that ERS estimates to be subject to some degree of development pressure (Figure 89). The total cost of preserving cropland subject to development pressure could be as much as $130 billion (Heimlich and Anderson, 2001).

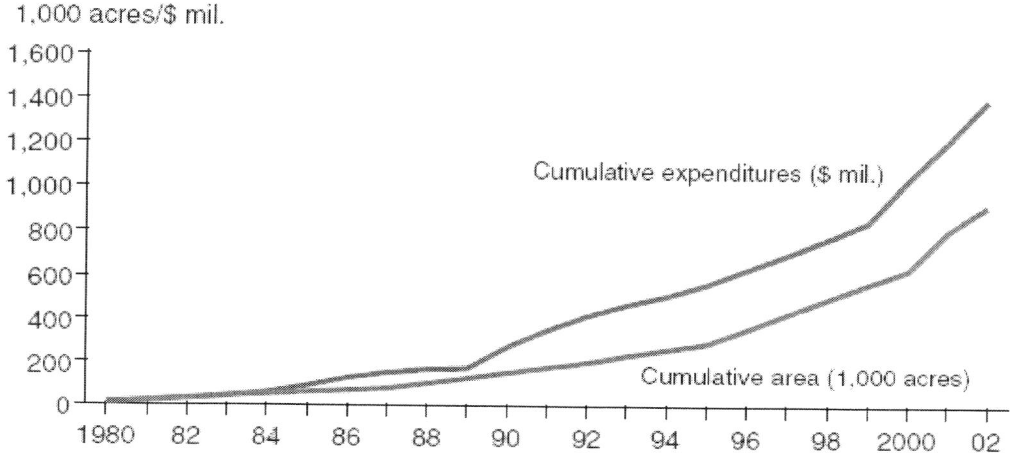

* PDR = Purchase of development rights.
Source: American Farmland Trust, 2004b. Data for some years are interpolated.

Figure 88. Accumulated expenditures and acreage in State PDR* programs are increasing.

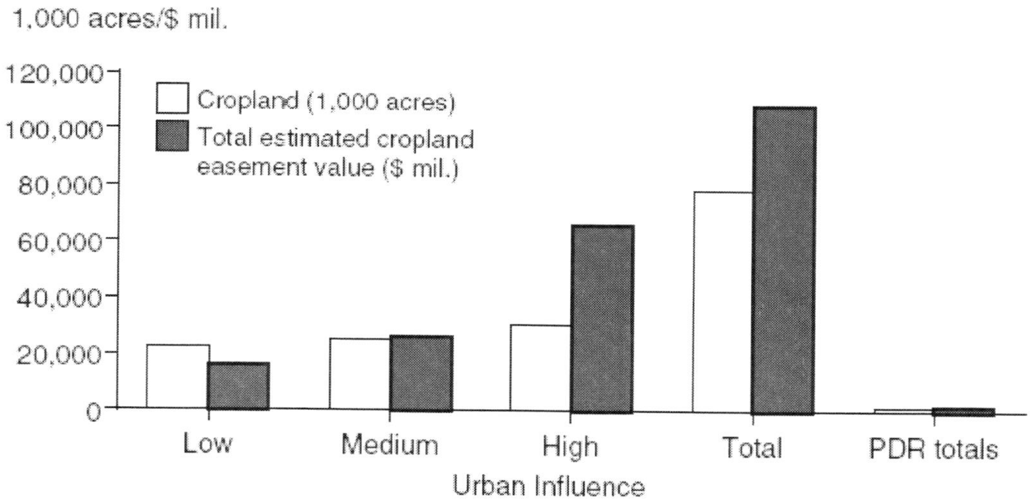

Notes: PDR totals represent cumulative funds spent and cumulative acres preserved by Federal, State, and local PDR programs through 2004 on all farmland types.
Source: ERS analysis of National Resources Inventory land use and NASS land values data (1994-1997); PDR totals from American Farmland Trust (2004b).

Figure 89. Cropland easement values and acres subject to urban influence versus actual PDR activity.

Despite State and local prerogatives in land use management, the Federal Government is increasingly partnering with local/State agencies and nonprofit organizations to protect farmland. Federal efforts to protect farmland began with the Agriculture and Food Act of 1981, which required Federal agencies to evaluate the impact of federally funded programs that converted farmland to nonagricultural uses and to consider alternative actions that would lessen the adverse impacts. Direct Federal involvement in permanent farmland protection did not begin until 1996, when the Farmland Protection Program (FPP) was established to help State, local, and tribal governments purchase agricultural conservation easements. The FPP distributed approximately $50 million during 1996–2001 in matching funds.

The 2002 Farm Security and Rural Investment Act reauthorized the FPP, which was renamed Farm and Ranch Land Protection Program (FRPP) through Executive rulemaking. FRPP provides up to 50 percent of easement costs on qualified, privately owned agricultural land. It also expanded the set of entities eligible to apply for funding to include nongovernmental organizations (primarily land trusts). Authorized funding increased to approximately $100 million per year for the 6 years beginning in 2002. With this increase, FRPP is now authorized to spend almost as much annually as all State PDR programs combined.

The high costs of permanently preserving farmland through PDR programs have generated support for TDR programs. While the sponsoring jurisdiction faces fewer costs, garnering taxpayer support in areas targeted to receive the urban densities being transferred is difficult, as is balancing the supply of and demand for development rights (Fulton et al., 2004). Fifty local jurisdictions have passed TDR ordinances, but only 15 TDR programs have individually preserved more than 100 acres (see AFT, 2001).

Many land trusts exist to preserve farmland (Figure 90). These private, nonprofit organizations accept donations of conservation easements on farmland and environmentally sensitive land. The donations benefit landowners in the form of Federal and State (in 10 States) income tax deductions. In Colorado, South Carolina, and Virginia, formal markets are developing that allow a landowner who donates an easement but cannot use the State tax credit to sell the unused credit to a third party (Conservation Fund, 2002).

ISSUES IN FARMLAND PROTECTION

The public benefits that are lost when farmland is converted cannot be readily measured in money terms. Instead, the benefits are typically estimated based on what people are willing to pay to avoid the losses associated with the conversion of farmland—i.e., the loss of agricultural production and rural amenities. Variation in local conditions leads to a wide range in estimates—from a fraction of a penny to more than a nickel per acre annually—to prevent the development of farmland. One analysis suggests this willingness to pay may exceed $1 billion annually for the United States (Heimlich and Anderson, 2001).

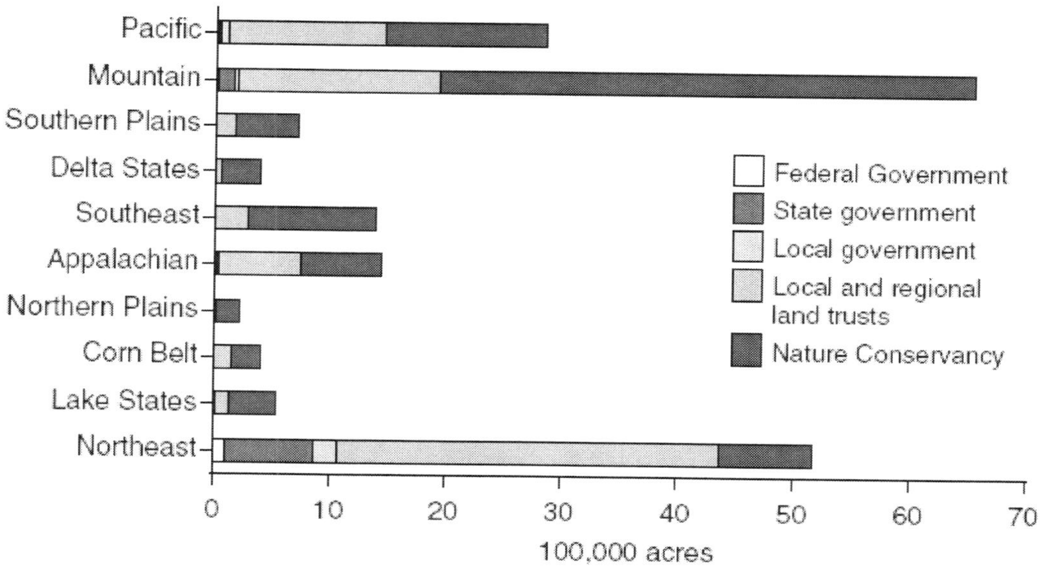

Notes: Nature Conservancy and Land Trust data include all acres protected, including farmland acres. Acres protected by local, State, and Federal governments are limited to farmland acres. May include some double counting of protected areas by different entities that collaborate to protect particular parcels of farmland. Nature Conservancy figures include purchases of fee-simple interests in land, in addition to acres on which only a conservation easement was purchased.

Source: Nature Conservancy data compiled by ERS from Wiebe, 1999; Local and regional trust data from Land Trust Alliance 2003 data (www.lta.org); farmland acres protected by State and local governments are from AFT (2004a,b) (www.farmlandinfo.org); farmland acres protected by Federal Government through Sept. 2004 are from data provided by NRCS, 2005.

Figure 90. Acres of land voluntarily protected, by sponsor.

Whether the benefits of farmland protection programs exceed program costs again depends heavily on local conditions. The direct costs of purchasing easements must be added to the value of urban benefits forgone when land is preserved (Lopez et al., 1994; Miller and Doering, 2004). Estimates for these opportunity costs are not readily available.

In addition to program costs, farmland protection programs have other impacts on government budgets and on resident taxpayers. Jurisdictions may save money on public service costs by preserving farmland because farmland requires fewer public services than residential uses do. Preserving land may benefit nearby residents who can look forward to rural scenic views and open space for the length of the easement (often into perpetuity). However, farmland preservation may impose costs on potential new residents who then have to live in higher densities elsewhere, face higher land prices, or endure longer commutes if they seek rural land farther from employment centers. How programs are implemented, and the distribution of enrolled lands, will determine the impacts on government budgets and taxpayers.

Farmland protection tools vary in their effectiveness at permanently preserving farmland, and providing intended benefits. For example, agricultural zoning exemptions allowing higher density residential development are common, and can limit the ability to preserve farmland. Agricultural districts may have limited success in areas where landowners commit to not develop only when their land faces little development pressure. Preferential assessment does little to preserve farmland in the long run because the capital gains from developing farmland usually exceed the rollback penalties for conversion. Preferential assessment may even encourage land speculation by reducing developers'costs of holding farmland in inventory.

Because they result in permanent (or at least 30-year) restrictions on non-farm development, PDR and TDR programs are considered to be the most effective in preserving agricultural lands. However, the actual effect of these programs on land development rates and patterns is uncertain. While the number of acres preserved can be counted, these programs may simply shift development pressures elsewhere. Also, compliance with/enforcement of development restrictions over the long term is not a sure thing.

An often-cited argument in support of PDR programs is that they help keep farmland affordable for new farmers. In theory, once the development rights have been sold, the market value of the preserved land will reflect only its value in a farming use, and may be significantly lower than its residential market value. However, a recent study found little evidence that easement restrictions significantly lowered preserved farmland prices (Nickerson and Lynch, 2001). It could be that landowners who farm as a recreational pursuit are outbidding "traditional" farmers for the land.

Though both TDR and PDR programs rely on conservation easements, economic implications and effectiveness can differ. Some PDR programs (due to ranking criteria and agency efforts to minimize costs) yield a pattern of preserved parcels that are widely separated. This raises questions about whether a "critical mass" of remaining farms can support farm input suppliers, and about the sustainability of remaining farms. TDR programs, on the other hand, have often been implemented in conjunction with reductions in allowed housing density (downzoning) of a large area. While many of the parcels in the downzoned area are not technically "preserved," the combination of zoning and TDRs may be effective at preventing widespread development. It is much more difficult to change zoning on an area-wide basis than on individual parcels. As a consequence, large clusters of "undeveloped" farmland (the downzoned area) may be preserved through TDR.

POLICY DEVELOPMENTS

Most recently, States have begun to implement "smart growth"strategies. Smart growth is a catchall phrase to describe a number of land use policies for influencing the pattern and density of new development. Without prohibiting development outside designated areas, smart growth policies use incentives and disincentives to direct new development to existing urban areas with appropriate infrastructure. PDR programs are one tool used to meet these goals. The effectiveness of smart growth will depend on how the incentive effects of new policies differ from pre-existing policies (Nickerson, 2001).

REFERENCES

[1] American Farmland Trust (2001). Fact Sheet: *Transfer of Development Rights*. Northampton, MA.

[2] American Farmland Trust (2004a). Fact Sheet: *Status of State PACE Programs*. Northampton, MA, July.

[3] American Farmland Trust (2004b). Fact Sheet: *Status of Local PACE Programs.* Northampton, MA.

[4] American Farmland Trust (1997). *Saving American Farmland: What Works.* Northampton, MA.

[5] Barnard, C.H. (2004). "Land Use Problems and Conflicts: The Interrelated Roles of Employment, Growth, Population Growth and Electronic Technologies," in J. Bergstrom, S. Goetz, and J. Shortle (eds.), *Land Use Problems and Conflicts*, Routledge, London.

[6] Barnard, C.H. (2000). "Urbanization Affects a Large Share of Farmland," *Rural Conditions and Trends* 10(2), U.S. Dept. Agr., Econ. Res. Serv.

[7] Conservation Fund (2002). "Buying and Selling Easement Tax Credits," *Common Ground: Conservation News from the Conservation Fund* 13(4): 4, Oct.-Dec.

[8] Fulton, W., J. Mazurek, R. Pruetz, and C. Williamson (2004). *TDRs and Other Market-Based Land Mechanisms: How They Work and Their Role in Shaping Metropolitan Growth.* Brookings Institution Center on Urban and Metropolitan Policy, June.

[9] Hellerstein, Daniel, Cynthia Nickerson, Peter Feather, Daniel Mullarkey, Abebayehu Tegene, Joseph Cooper, Dwight Gadsby, and Charles Barnard (2002). *Farmland Protection: The Role of Public Preferences for Rural Amenities.* AER-815. U.S. Dept. Agr., Econ. Res. Serv.

[10] Heimlich, Ralph, and William D. Anderson (2001). *Development at the Urban Fringe and Beyond: Impacts on Agriculture and Rural Land.* AER-803. U.S. Dept. Agr., Econ. Res. Serv.

[11] Lopez, R.A., F.A. Shah, and M.A. Altobello (1994). "Amenity Benefits and the Optimal Allocation of Land," *Land Economics* 70: 346-58 (Feb).

[12] Miller, Douglas, and Otto Doering (2004). "Purchase of Development Rights Programs: Assessments of Their Impact on Development Trends and Net Benefits." Unpublished manuscript, Purdue University.

[13] Nickerson, Cynthia (2001). "Smart Growth: Implications for Agriculture in Urban Fringe Areas," *Agricultural Outlook* (April), U.S. Dept. Agr., Econ. Res. Serv..

[14] Nickerson, Cynthia J., and Lori Lynch (2001). "The Effect of Farmland Preservation Programs on Farmland Prices," *American Journal of Agricultural Economics* 83(2):341-351 (May).

[15] Vesterby, Marlow, and Kenneth S. Krupa (2001). *Major Uses of Land in the United States, 1997.* SB-973. U.S. Dept. Agr., Econ. Res. Serv.

[16] Vesterby, Marlow, and Kenneth S. Krupa (1997). *Major Land Uses* (data, 1945-97).

In: Agricultural Resources and Environmental Indicators ISBN 978-1-60021-467-7
Editor: Keith Wiebe and Noel Gollehon, pp. 253-259

Chapter 27

FEDERAL LAWS PROTECTING ENVIRONMENTAL QUALITY

Marc Ribaudo

Federal environmental laws can influence farmers'decisions about production practices or input use. These laws use a variety of mechanisms for protecting the environment, ranging from voluntary incentives to regulatory approaches.

INTRODUCTION

Farmers face a complex set of factors when they make decisions about farm management and conservation practices. The vagaries of weather and markets introduce uncertainty into farmers'operations. The use of conservation practices may also introduce uncertainty about net returns while producing benefits enjoyed mostly off the farm. Decisions made by farmers on how and where to produce commodities can be influenced by policies and programs for protecting the environment. USDA has several major programs for providing financial and technical assistance to farmers for protecting water quality, soil quality, and wildlife habitat (see AREI Chapter 5).

Farmers can also be influenced by other Federal environmental protection policies and programs that may restrict certain production practices. This chapter will focus on these programs. The U.S. Environmental Protection Agency is chiefly responsible for administering these policies and programs. The **Clean Water Act** (1972) is the major law protecting water quality. Several CWA programs address "nonpoint-source" pollution, which is the most prevalent type of pollution associated with agriculture (see Chapter 2, "Water Quality Impacts of Agriculture").

The Nonpoint Source Program (Section 319) requires States to develop nonpoint-source management programs. Nonpoint-source control plans can include State regulatory measures, but usually emphasize voluntary actions like those used in USDA conservation programs. Implementation grants to States and tribes ($200 million in FY2005) fund projects like installation of best management practices (BMPs) for animal waste; design and implementation of BMP systems for stream, lake, and estuary watersheds; and basin-wide landowner education programs. The Clean Water State Revolving Fund (CWSRF), created by Congress to fund the construction of water treatment plants, can be used by States to provide reduced-rate loans for water quality projects included in the State nonpoint-source plan. Fifteen States have used CWSRF for funding waste management systems, manure spreaders, conservation tillage equipment, irrigation equipment, filter strips, and streambank stabilization.

Water pollution from some animal feeding operations is treated as a point source under the Clean Water Act (see Chapter 4, "Animal Agriculture and the Environment"). Confined animal feeding operations meeting certain size thresholds or other conditions fall under the National Pollution Discharge Elimination System (NPDES, Section 402). These operations, known as Concentrated Animal Feeding Operations (CAFOs), must obtain NPDES permits that specify standards for the production area (i.e., housing, waste storage) and for the land where wastes are applied. CAFOs must also implement a nutrient management plan for animal manure applied to land, a significant change for Federal water quality laws.

As a form of nonpoint pollution, nutrient runoff from fields has traditionally been addressed with voluntary approaches. This is the first time that a nonpoint source of water pollution has been regulated at the Federal level. EPA estimates that up to 15,500 operations are covered by the CAFO regulations. These regulations may impose significant manure management costs in areas where land for spreading manure is scarce. These costs could influence location decisions for large operations and spur the development of alternative uses for manure. EPA encourages CAFOs to seek financial and technical assistance from USDA to help them meet manure management requirements.

The **Total Maximum Daily Load** provisions of the Clean Water Act are intended to be the second line of defense for protecting the quality of surface water resources. When technology-based controls are inadequate for water to meet State quality standards, Section 303(d) of the Clean Water Act requires States to identify those waters and to develop total maximum daily loads (TMDL). A TMDL is the maximum amount of a pollutant that a water body can receive and still meet water quality stan-dards, and an allocation of that amount to all the pollutant's sources. States must submit to EPA a list of impaired waters and the cause of the impairment. More than 20,000 such waters have been identified as impaired under Section 303(d) (Figure 91).

Among the top impairments are sediment, nutrients, and pathogens. States, territories, and authorized tribes are responsible for establishing a program to meet TMDLs. Point-source reductions to meet wasteload allocations are achieved through NPDES permits. Agricultural nonpoint sources are generally addressed through voluntary programs, but States may use regulations. TMDLs have not generally been used to regulate agricultural production, but particular management practices are required on agricultural operations in three TMDL-designated watersheds in California.

Section 404 of the Clean Water Act establishes a program for protecting wetlands. It regulates the discharge of dredged and fill material into U.S. waters, including wetlands, and is a key policy for meeting the "no net loss" goal for wetland acreage. Section 404 contains a review process that handles small conversions through general permits. More thorough, qualitative reviews are conducted for major proposals affecting wetlands. Activities regulated under this program include fills for development, water resource projects (such as dams and levees), infrastructure development (such as highways and airports), and conversion of wetlands to uplands for farming and forestry. Under the law, a permit is required to fill a wetland, and is granted only if impacts to wetlands are minimized. Compensation for any unavoidable impacts is made through wetland restoration elsewhere. The U.S. Army Corps of Engineers administers the program, while EPA develops and interprets environmental criteria used in evaluating permit applications. Ongoing farming activities are generally exempt from Section 404, but filling wetlands to create new farmland would require a permit.

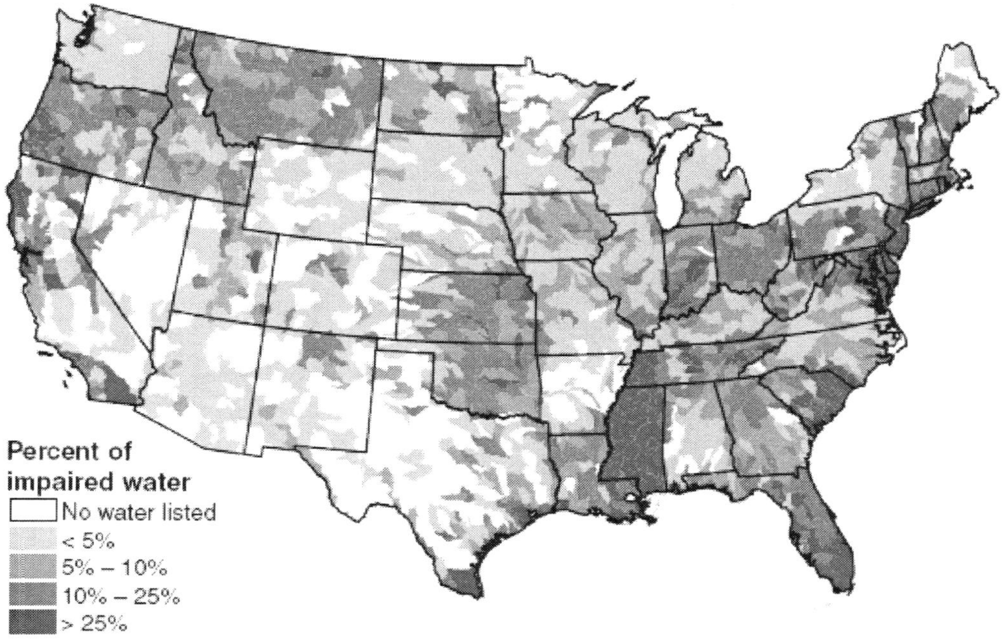

Source: U.S. EPA.

Figure 91. Impaired watersheds, 2000.

The Coastal Zone Management Act Reauthorization Amendments (CZARA) of 1990 added nonpoint-source water pollution requirements to the Coastal Zone Management Act of 1972. CZARA requires that each State and territory with an approved coastal zone management program submit to EPA and to the National Oceanic and Atmospheric Administration a program to implement management measures for nonpoint-source pollution to restore and protect coastal waters. A list of economically achievable management measures for controlling agricultural nonpoint-source pollution is part of each State's management plan. States can initially use voluntary incentive mechanisms such as education, technical assistance, and financial assistance, but may enforce management measures if voluntary approaches fail. Currently, 34 coastal States and territories have developed nonpoint-source pollution control plans (Figure 92).

The **Safe Drinking Water Act** (SDWA) of 1974 requires the EPA to set standards for drinking-water quality and requirements for water treatment by public water systems. States are required to develop Source Water Assessment Programs to assess the areas serving as public sources of drinking water in order to identify potential threats and to initiate protection efforts. Each assessment must include four elements: (1) delineating (or mapping) the source water assessment area, (2) conducting an inventory of potential sources of contamination in the delineated areas, (3) determining the susceptibility of the water supply to those contamination sources, and (4) releasing the results of the determinations to the public. Under the 1996 amendments, EPA is required to establish a list of contaminants for consideration in future regulation. The Drinking Water Contaminant Candidate List, released in March 1998, lists chemicals by priority for (a) regulatory determination, (b) research, and (c) monitoring. Several agricultural chemicals—including metolachlor, metribuzin, and the triazines—are among those to be considered for potential regulatory action. Also under the 1996 amendments, water suppliers are required to inform their customers about the levels of certain contaminants (and associated EPA standards), and the likely sources of the contaminants.

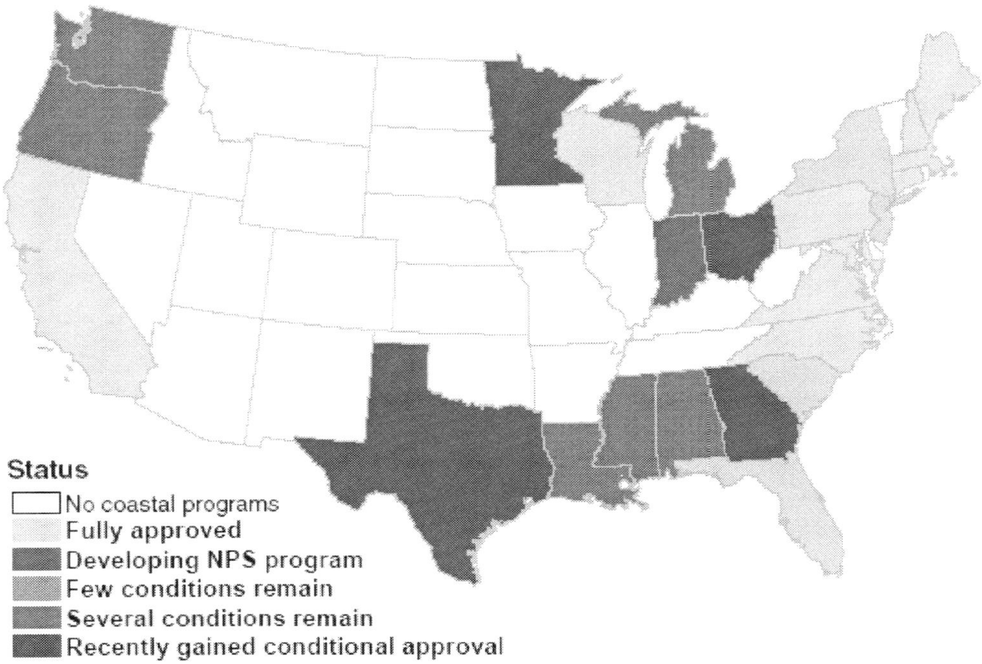

Source: U.S. Department of Commerce, NOAA.

Figure 92. Coastal nonpoint-source pollution control programs, 2005.

The **Clean Air Act** (CAA) of 1970 sets limits on how much of a pollutant can be in the air anywhere in the United States. Under Section 110, each State must develop a State Implementation Plan (SIP) to identify the sources of air pollution and to determine what reductions are required to meet Federal air quality standards. A SIP is a collection of the regulations a State will use to clean up polluted areas. Pollutants regulated under the CAA are called criteria air pollutants. Permissible emission levels are generally based on health concerns, but visibility standards may also apply. The criteria pollutant most associated with agriculture is particulates. Where airborne dust from fields, burning crop residues, or other sources exceeds permissible levels, States must take steps to reduce emissions. Airborne dust from fields in Washington and particulates from burning rice straw in California have led to SIPs for controlling emissions from agricultural fields.

Ammonia is a precursor for fine particulates in the atmosphere, and confined animal operations are the source for over 70 percent of ammonia emissions in the United States.

California has implemented State regulations for reducing ammonia emissions from dairy operations that were affecting air quality in heavily populated areas downwind. EPA recently revised the particulate matter standard to control for fine particulates. This could result in States' requiring animal feeding operations to control ammonia emissions.

The **Federal Insecticide, Fungicide, and Rodenticide Act** (FIFRA) of 1947 provides direct controls over the sale and use of pesticides. Under FIFRA, all pesticides must be approved by EPA through a mandatory registration process. Products determined to pose an unacceptable risk to human health or to the environment can be denied registration, thereby preventing their distribution and use. Fifty pesticides and pesticide formulations have been banned under FIFRA as of 2004.

In 1996, the **Food Quality Protection Act** amended FIFRA to eliminate inconsistencies between it and the Federal Food, Drug, and Cosmetic Act of 1938 (which regulates pesticide residues on food). The amendments allow EPA to move quickly to suspend the use of a pesticide to prevent serious risks to human health and the environment. The amendments also provide incentives for the development and maintenance of minor use registrations. Minor uses of pesticides are defined as uses for which pesticide product sales do not justify the costs of developing and maintaining EPA registrations. Lack of registrations can limit the pest control tools available to the growers of "minor" crops (including many fruits and vegetables).

The **Endangered Species Act** (ESA) of 1973 conserves the ecosystems upon which endangered and threatened species (wildlife and plants) depend. To do so, the law regulates the modification or degradation of habitat deemed critical for species survival. All Federal agencies are required to protect endangered species and protect their habitat. Private landowners who wish to conduct activities on their land that might incidentally harm wildlife listed as endangered or threatened are required to obtain an incidental take permit from the U.S. Fish and Wildlife Service. To obtain a permit, the landowner must develop a Habitat Conservation Plan. The plan is designed to offset any harmful effects the proposed activity might have on the species. Under the Act, EPA must also ensure that the use of pesticides it registers will not result in harm to any species listed as endangered and threatened by the U.S. Fish and Wildlife Service, or to habitat critical to those species'survival. EPA's Office of Pesticide Programs initiated the voluntary Endangered Species Protection Program in 1988 to protect endangered and threatened species from harm due to pesticide use. Labels of certain pesticides contain information to help users minimize the risk of pesticide use in critical habitat areas. At least 1 county in 24 States has pesticide use restrictions under this program (Figure 93).

The Endangered Species Act may have a large impact on agriculture through the supply of irrigation water from Federal irrigation projects. The Bureau of Reclamation has taken measures to protect the flow of rivers supporting endangered species, such as salmon. Sufficient flow for endangered species can reduce the irrigation water available to farmers from Federal irrigation projects, with obvious implications for crop production.

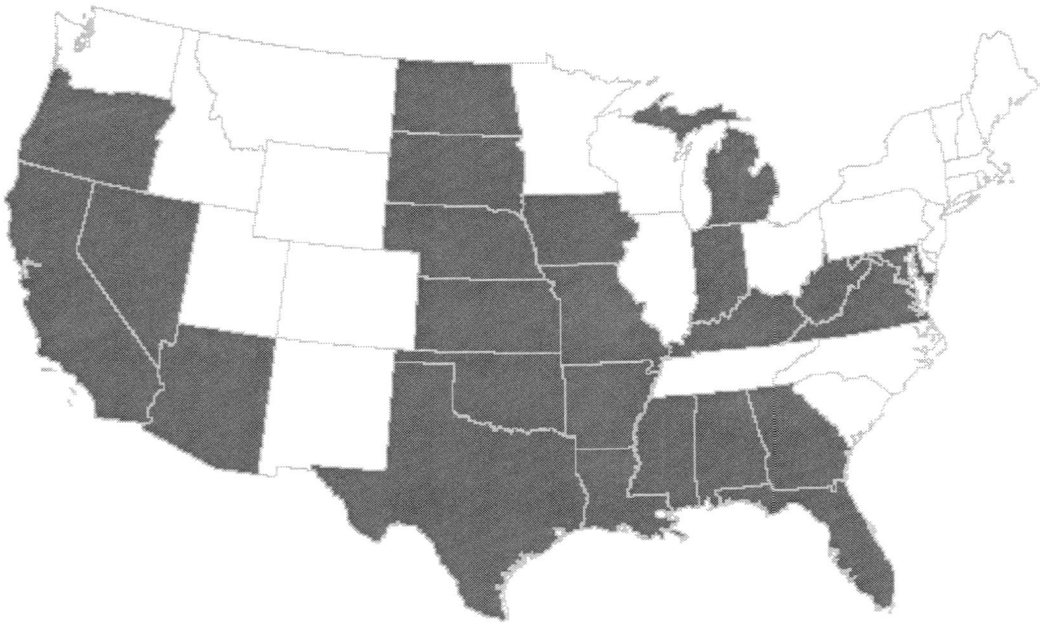

Source: U.S. EPA.

Figure 93. States with pesticide use limitations under the Endangered Species Protection Program, 2005.

For example, the ESA triggered a complete shutdown of irrigation water to more than 1,300 farms and ranches in the Klamath River Basin during a drought in the spring of 2001. Conflicts over the ESA's implementation in the irrigated West will continue to be the source of many legal actions.

The Comprehensive Environmental Response, Compensation, and Liability Act (CERCLA), commonly known as Superfund, was enacted in 1980 to provide broad Federal authority to respond to releases of hazardous substances that might endanger public health. CERCLA requires reporting to EPA when a facility releases more than a "reportable quantity" (100 pounds in a 24-hour period) of a hazardous substance. EPA is authorized to require long-term remedial action that permanently and significantly reduces threats to public health. Originally focused on hazardous wastes from industrial plants, the increased size and consolidation of animal feeding operations has raised the possibility that the emission of substances like ammonia and hydrogen sulfide from such operations may be subject to the notification provisions of CERCLA (U.S. EPA, 2005).

Role of USDA Conservation Programs

Federal environmental laws cover many aspects of agricultural production. Laws aimed at preserving habitat (Section 404 of the Clean Water Act, Endangered Species Act) or at controlling the use of toxic agricultural inputs (FIFRA) are the source of direct constraints on agriculture at the Federal level. Those Federal laws directed at reducing pollution to the environment (i.e., Clean Water Act, Clean Air Act, Coastal Zone Management Act) have generally not constrained agriculture directly, opting instead for voluntary approaches overseen primarily by the States. Constraints on agricultural production to reduce pollution emissions are more likely to arise at the State level in response to local problems.

USDA's conservation programs can help farmers respond to resource issues subject to regulation. For example, being in a 303(d) impaired watershed is a screening advantage in applications for EQIP (see Chapter 5.4, "Working-Land Conservation Programs"). EPA encourages CAFOs to seek financial and technical assistance from USDA to help them implement Clean Water Act provisions, and 60 percent of EQIP's funding is earmarked for animal feeding operations. USDA also helps farmers reduce air pollution in dust and ozone nonattainment areas in California with a cost-share program funded through EQIP. The Wildlife Habitat Incentives Program is being used to help landowners protect habitat for endangered species. The Conservation Reserve Program (see Chapter 22, "Land Retirement Programs") and Grassland Reserve Program both consider potential benefits to endangered species in the selection of land offered for enrollment.

REFERENCES

[1] U.S. Environmental Protection Agency (2005). *Animal Feeding Operations Air Quality Compliance Agreement Fact Sheet.* Available at www.epa.gov/compliance/resources/agreements/caa/cafo-fcsht-0501.html

In: Agricultural Resources and Environmental Indicators ISBN 978-1-60021-467-7
Editor: Keith Wiebe and Noel Gollehon, pp. 261-266

APPENDIX: USDA RESOURCE SURVEYS AND DATA INVENTORY

Robert Dubman

USDA maintains databases that help enhance knowledge about the Nation's agricultural resources. This appendix inventories the major databases, including many used in AREI. Most of these databases are confidential, with only summarized estimates publicly available. Special access can be granted for qualified researchers.

AGRICULTURAL RESOURCE MANAGEMENT SURVEY (ARMS)

The annual Agricultural Resource Management Survey (ARMS) is USDA's primary source of information on the production practices, resource use, financial condition, and economic well-being of U.S. farm households. Summarized ARMS estimates are available on the ERS website in a dynamic, technologically advanced, and easy-to-use web-based delivery tool. The four major areas covered are Crop Production Practices, Commodity Costs of Production, Farm Business Structure and Finance, and Farm Households. Starting with the 2003 ARMS, a greatly expanded sample allows detailed data analysis of the top 15 agricultural producing States. An online ARMS briefing room houses the latest ARMS-based publications and estimates.

Estimates from ARMS data are essential to USDA, congressional, administration, and industry decisionmakers when weighing alternative policies and programs that touch the farm sector or affect farm families. Sponsored jointly by ERS and NASS, ARMS is the only national survey that provides observations of commodity-specific, field-level farm practices; the economics of the farm business operating the field (or dairy herd, poultry house, etc.); and the characteristics of the U.S. farm household.

ARMS data underpin USDA's annual estimates of net farm income, subsequently provided to the Bureau of Economic Analysis for estimating gross domestic product and personal income. ARMS fulfills a congressional mandate that USDA provide annual cost-of-production estimates for commodities covered under farm support legislation. ARMS also provides data regarding chemical use on field crops as required under environmental and food safety legislation.

ARMS is conducted in three phases each year. A screening phase, in June-August, collects general farm data on crops grown, livestock produced, and farm sales. These data are used to identify farms to be contacted for Phases II and III. Phase II, conducted in October-December, collects data associated with agricultural production practices, resource and input use, and production. Phase III, in February-April, gathers data on farm income, expenditures, and cost of production for specific commodities and on the financial condition of farms. The ARMS is conducted mostly by personal enumeration of farmers—a self-enumerated mail-in version started in 2003 for the expanded sampling in the top 15 agricultural States. A complex multiframe, stratified sampling procedure is applied. The results are weighted and aggregated to develop State, regional, and national estimates.

ARMS Phase II

Crop Production Practices and predecessor surveys were conducted annually from 1964 through 2004 by USDA's NASS with funding from ERS. In 1996, the annual Cropping Practices Survey was merged into ARMS. Phase II of ARMS is USDA's primary source of information about the current status and trends in crop production practices for several large-acreage field crops (corn, soybeans, wheat, and cotton). This phase also obtains data on U.S. farmers' agricultural resource use, as well as data to assess potential environmental impacts of crop production practices.

Tailored Phase II reports going back to 1996 are available on the ERS ARMS web tool. ARMS Phase II gathers data from randomly selected acres of a specified crop. Farm operators are asked to provide field-level information on all fertilizer, pest, nutrient, and crop residue treatments, all tillage operations prior to planting, and data on other inputs and cultural practices. Data can be summarized by crop, year, ERS Farm Resource Region, irrigation system, previous crop, and tillage system. The operator also identifies whether the field had been designated as highly erodible land (HEL) by USDA's Natural Resources Conservation Service and whether the farm participated in farm price and income support programs. All Phase II respondents are asked to complete a Phase III farm and household financial survey linking cropping practices to financial performance.

The Farm Costs and Returns Survey (FCRS) was the main precursor to ARMS and was conducted annually from 1985 to 1995, with funding and support from NASS and ERS. ARMS was developed by combining the former Cropping Practices Survey (CPS), the Farm Costs and Returns Survey (FCRS), and the commodity cost of production surveys. The FCRS was conducted to gather information on the financial situation of farm and ranch businesses, the costs of producing various crop and livestock commodities, and the characteristics and financial situations of farm operators and their households.

Chemical Use Surveys

Chemical Use Surveys were initially funded under the 1989 President's Food Safety Initiative. Fruit and vegetable crops are the primary target of the survey program, with even-year surveys to cover vegetables and odd-year surveys to cover fruits. In each year, certain commodities are targeted to obtain more comprehensive information on management practices and costs, with recent emphasis on Integrated Pest Management and organic production. The surveys are conducted by NASS using personal enumeration of a stratified systematic sample of commercial growers. The surveys have gathered data on pesticide use for most commercial production of fruits and vegetables in the United States (see AREI Chapter 4).

Census of Agriculture

The Census of Agriculture is conducted every 5 years, with the most recent in 2002. In 1996, responsibility for the Census of Agriculture was transferred from the U.S. Department of Commerce to USDA's National Agricultural Statistics Service (NASS). The Census attempts to be a complete enumeration of the general characteristics of all agricultural operations. However, it uses a random sampling procedure to estimate a wide variety of financial and operator characteristics. The Census of Agriculture and the ARMS survey overlap in census years. In these years, ARMS questionnaires are adjusted so that a farmer responding to the ARMS meets all the obligations of the Census.

Farm and Ranch Irrigation Survey (FRIS)

FRIS is a follow-on survey to the Census of Agriculture. FRIS provides data about irrigated agriculture by State and by Water Resource Area. All producers who report irrigation in the Census are eligible to receive a FRIS questionnaire, though the survey does not include irrigation on horticultural specialty, institutional, experimental, research, and Indian reservation farms. Data were collected in 1979, 1984, 1988, 1994, 1998, and 2003. Responsibility for FRIS and the Census of Agriculture was transferred to USDA from the Department of Commerce's Bureau of Census starting in 1997. The survey is based on a stratified, random sample of irrigators, and then adjusted to represent all eligible irrigators. The FRIS data are collected to be statistically reliable for the conterminous United States and within each of the 18 major Water Resource Areas. Data are collected on irrigation water sources, costs, energy use, maintenance of equipment, application technologies and frequency, crop yields, water conservation activities, and water management practices (see Chapter 4).

National Resources Inventory (NRI)

NRI is a statistical survey of natural resource conditions, land use, and trends on nonfederal land. The NRI was conducted by USDA-NRCS field staff every 5 years during 1977-97, but is now conducted annually. Transition to a fully implemented annual NRI is taking place over several years. Information is collected on the status, condition, and trends of land, soil, water, and other resources on the Nation's land (including all States and territories except Alaska). Data for the 2003 NRI were collected from more than 800,000 sample locations and are statistically reliable for national, regional, State, and substate analysis. The 2003 NRI provided a nationally consistent data base that was constructed specifically to estimate trends for natural resources from 1982 to 2003 (see AREI Chapters 1, 2, and 4).

CONSERVATION EFFECTS ASSESSMENT PROJECT (CEAP)

CEAP is a USDA effort conducted by NRCS and NASS designed to assess the environmental effects of the 2002 Farm Security Act conservation programs. It is based on NRI sample points and examines nutrient, manure, pest management, buffer system, tillage, irrigation, and drainage practices as well as wetland protection and restoration. CEAP provides a link between farm production choices and NRI environmental data. An additional pilot survey integrated the ARMS and CEAP surveys for wheat farms into a single instrument, CEAP-ARMS, for the 2004 and 2005 calendar years. For the first time, an integrated USDA survey will allow data links between operator household/farm financial characteristics and farmers'conservation practice/environmental performance data.

CONSERVATION COMPLIANCE STATUS REVIEW

USDA's Natural Resources Conservation Service conducts status reviews of tracts determined to be highly erodible land (HEL), using a 3-percent random sample. The sample is statistically reliable at the State level for States with large HEL acreage and with high participation in USDA programs. Each tract in the sample was visited to determine the extent of compliance with the HEL provisions of the 1985 and subsequent Farm Acts. The review results were aggregated to State, regional, and national estimates, housed in the Compliance Reviews Database System. In 2000, the FSA data collection process was revamped to provide a nationally uniform means of collecting, maintaining, analyzing, and reporting compliance review data (see Chapter 5)

Conservation Reserve Program (CRP) Contract Data

USDA's Farm Service Agency (FSA) develops and maintains data on all tracts enrolled in the CRP, based on information provided by program participants and observations by FSA during onsite inspections. This data set includes information on the type of contract, location, acreage enrolled, land capability class and subclass, rental rate paid, average soil-specific rental rate, and cost sharing (see Chapter 5).

CROP RESIDUE MANAGEMENT (CRM) SURVEY

The CRM survey was conducted by the Conservation Technology Information Center (CTIC) in 1998, 2000, 2002, and 2004 to provide State and national statistics on various conservation tillage systems. CTIC is a division of the National Association of Conservation Districts and is administered by industry, government agencies, commodity organizations, and growers. The CRM survey provides estimates on five different tillage systems for field crops: no-till, mulch till, ridge till, reduced till (15-30 percent residue), and conventional till (less than 15 percent residue). Local directors of USDA program agencies and others knowledgeable about local residue management practices complete the survey each summer as a group effort. These local judgments are summarized to provide State, regional, and national estimates. In addition, several States conduct statistically derived transects to survey crop residue levels (see Chapter 4).

CURRENT RESEARCH INFORMATION SYSTEM (CRIS)

CRIS, a research information database, maintains data on all agricultural and forestry research funded by USDA, including research by problem area, subject, field of science, funding, objectives, approach, performing organizations, and responsible individuals. USDA's Agricultural Research Service (ARS) maintains the system (see Chapter 3).

JUNE AGRICULTURAL SURVEY

The largest single sample-based survey NASS conducts each year is the multiple-frame June Agricultural Survey (JAS). The JAS and other minor annual surveys focus on agricultural production for major crops, livestock, and associated inventories. These surveys collect farm-level data to produce State and U.S. crop forecasts and estimates published in the NASS Agricultural Statistics Board reports. NASS produces approximately 400 reports each year, with information released on a weekly, monthly, quarterly, or annual basis depending on the commodity. The Agricultural Land Values Survey (ALVS) was combined into this NASS series in 1994 when questions on land values and cash rents were added (see Chapter 1).

AREA STUDIES PROJECT

USDA's Area Studies Project was a trial survey designed to characterize the extent of adoption of nutrient, pest, soil, and water management practices and to assess the factors that affect adoption for a wide range of management strategies across different natural resource regions. A detailed field-level survey was administered to farmers in 12 watersheds to gather data on agricultural practices, input use, and natural resource characteristics associated with farming activities. Surveys conducted in each area between 1991 and 1993 collected detailed information on production technologies, cropping systems, and agricultural practices at both the field and whole-farm level. The survey sample points corresponded with National Resource Inventory (NRI) sample points, for which NRCS had collected soil, water, and other natural resource data. Recent CEAP-ARMS and ARMS supplanted these data (see AREI Chapters 4).

OTHER DATA

Data on real property taxes (State and local) on farm and ranch lands and buildings were collected annually through a nationwide mail survey of over 4,000 taxing officials until the survey was discontinued in 1995. The survey, conducted by ERS, provided tax and acreage information on about 42,000 parcels of farm and ranch lands in the 48 contiguous States. Internal Revenue Service databases of taxpayers that file Schedule F provide other tax information.

Data on foreign ownership of U.S. agricultural land are collected under the auspices of the Agricultural Foreign Investment Disclosure Act of 1978 (AFIDA). This act requires all foreign owners of U.S. agricultural land to report their holdings to the Secretary of Agriculture. Acquisitions and dispositions of such land by foreign owners are to be reported as they occur. This provides USDA with a continuing inventory of such ownership, which is netted out at the end of each calendar year and reported to the President and Congress.

Cropping Practices Survey— see ARMS. Farm Costs and Returns Survey— see ARMS. Agricultural Land Values Survey— see June Agricultural Survey.

INDEX

A

B

D

E

F

G

H

I

J

K

L

M

N

O

P

T

U

V

W

Y